Assistant
Systems Engineering Dept
U. of Petroleum & Minerals
Dhahran, Saudi Arabia

PROJECT MANAGEMENT
with CPM and PERT

PROJECT MANAGEMENT
with CPM and PERT
SECOND EDITION

Joseph J. Moder

Professor and Chairman, Department of Industrial Engineering and Systems Analysis, University of Miami

Cecil R. Phillips

Manager, Textile Systems, Kurt Salmon Associates, Inc.

VNR

VAN NOSTRAND REINHOLD COMPANY
NEW YORK   CINCINNATI   TORONTO   LONDON   MELBOURNE

Van Nostrand Reinhold Company Regional Offices:
New York   Cincinnati   Chicago   Millbrae   Dallas

Van Nostrand Reinhold Company International Offices:
London   Toronto   Melbourne

Manufactured in the United States of America.

Published by Van Nostrand Reinhold Company
450 West 33rd Street, New York, N.Y. 10001

Published simultaneously in Canada by
Van Nostrand Reinhold Ltd.

15   14   13   12   11   10   9   8   7   6   5

WE DEDICATE THIS BOOK

to the eleven other Moders and Phillipses who patiently did without their fathers and husbands for too many evenings and weekends.

PREFACE TO
SECOND EDITION

The year 1969 marked the end of a decade since the first publications appeared on the Critical Path Method (CPM) and the Program Evaluation and Review Technique (PERT). These ten years have seen a continuation of the rapid growth in the technology, both in practical applications and in further development of the theory. Although it seemed clear in 1964—when the First Edition of this text was published— that the network approach was a substantial contribution to management methods, it has now become established as a basic tool in the field.

The applications of critical path methodology have expanded to virtually every type of activity that could be considered a "project"—from surgical procedures to the complete construction of World Fairs, from the development of a new product to the pilot production of several units. It has also been applied to the detailed analysis of repetitive tasks such as the month-end closing of books. The computer programs available for

CPM and PERT processing have continued to increase in number and in capabilities. There is also evidence that the methodology is being applied widely throughout the world. This text alone has enjoyed a substantial international distribution and has been translated into other languages, including Russian and Chinese.

Significant advancements have also been made in the theoretical aspects. Scholars have found that the network is a graphical vehicle of considerable utility in describing and formulating a variety of problem types. For example, the problems of scheduling job shops are increasingly being described as network resource allocation problems. In this and other areas of industrial management and operations research, graduate students have found that critical path technology provides a rich hunting ground for thesis topics.

In revising this text we have attempted to bring it up to date by incorporating the most meaningful developments in both theory and practice. These developments, as expected, have tended to eliminate the practical distinctions between the original CPM and PERT techniques, evolving instead a body of knowledge that is common to both. Unfortunately, however, no single term has been accepted that describes the common discipline, and we feel obliged to continue using the double name "CPM and PERT," along with "network methods," "critical path methods" (the term we prefer), and others.

The organization of the chapters remains essentially the same as in the first edition, except that the basic and advanced subjects have been more clearly identified by grouping the subjects into Parts I and II, respectively. The five chapters of Part I comprise a complete course in the fundamentals of the planning and scheduling features of critical path methods, including manual and computer methods of calculation. One consistent set of terms and symbols is used throughout Part I. (Those familiar with the First Edition will note that the Space Symbols have been simplified, so that a special template is not required.) Thus, industrial and commercial users of the methods may study only Part I in preparation for most practical applications. A good two or three-day training course for industrial personnel can be based on Part I, with selected portions from Part II as appropriate to the needs of the group.

For more advanced users and college students, Part I may serve as an introduction to the more specialized topics of Part II. Beginning with a new Chapter 6, for example, the reader is exposed to the variety of networking schemes that have developed over the decade, and he may choose to adopt one other than the basic arrow scheme employed in Part I. Chapter 7 is also new and covers some of the essential elements of critical path processing from the point of view of the computer programmer.

The other chapters have been updated to include the recent advances in the respective topics. Resource allocation, in particular, has received heavy attention in the technical literature, and our chapter on this subject has been almost completely rewritten. The material in Part II is suitable for college level courses in departments of industrial engineering and management, business administration, civil engineering, and systems engineering. While certain portions of the text require some mathematical statistics and linear programming background, the prerequisites generally are satisfied by upper level undergraduate students.

Of interest to college instructors will be the added exercises at the end of most chapters, and perhaps more important, the inclusion of many of the problem solutions at the end of the book.

In closing, we would like to express our sincere thanks to Mary Dutcher, whose ability to transform the hand written page is phenomenal.

Joseph J. Moder
Coral Gables, Florida

Cecil R. Phillips
Atlanta, Georgia

April 1970

CONTENTS

NOMENCLATURE

$a =$ the "optimistic" performance time estimate used in PERT—the time which would be bettered only one time in twenty, i.e., the fifth percentile (where specifically noted, it will also be used to denote the zero percentile used in conventional PERT).

$b =$ the "pessimistic" performance time estimate used in PERT—the time which would be exceeded only one time in twenty, i.e., the ninety-fifth percentile (where specifically noted, it will also be used to denote the 100 percentile used in conventional PERT).

$C_d =$ direct costs associated with the performance of an activity in time d, the "crash" performance time.

$C_D =$ direct costs associated with the performance of an activity in time D, the "normal" performance time.

$d =$ "crash" activity performance time—the minimum time in which the activity can be performed.

$D =$ "normal" activity performance time—the one which minimizes the activity direct costs; also used to denote the mean activity performance time based on a single time estimate.

$\Delta c =$ incremental reduction in the direct costs of performing an activity, used in time-cost trade-off studies.

$\Delta t =$ incremental reduction in the time required to perform an activity, used in time-cost trade-off studies.

$E =$ earliest (expected) event occurrence time.

$ES =$ earliest (expected) activity start time.

$EF =$ earliest (expected) activity finish time.

$FS =$ activity free slack (or float).

$L =$ latest allowable event occurrence time.

$LS =$ latest (expected) activity start time.

$LF =$ latest (expected) activity finish time.

$m =$ the "most likely" performance time estimate used in PERT—the modal value of the performance time distribution.

$S =$ total activity slack (or float).

$t =$ actual activity performance time, determined after the activity has actually been completed.

$t_e =$ mean activity performance time based on the three (PERT) time estimates, a, m, and b.

$T =$ actual occurrence time of a specific network event, determined after the event has actually occurred.

$T_d =$ total (expected) project duration time achieved by using "crash" activity performance times on all critical path activities.

$T_D =$ total (expected) project duration time achieved by using all "normal" activity performance times.

$T_s =$ scheduled event occurrence time.

$V_t =$ the estimated variance of the actual activity performance time, t, based on the PERT formula $[(b - a)/3.2]^2$.

$(V_t)^{1/2} =$ square root of V_t, called the standard deviation of the actual activity performance time, t.

$V_T =$ the estimated variance of the actual occurrence time, T, of a specific network event.

$(V_T)^{1/2} =$ square root of V_T, called the standard deviation of the actual event occurrence time, T.

$Z =$ standard normal deviate, equal to the difference between a random variable, such as T, and its expected or scheduled time, such as T_s, divided by the standard deviation of the random variable, such as $(V_T)^{1/2}$.

I
BASIC TOPICS

1

INTRODUCTION

Project management will be interpreted quite broadly in this text in order to encompass the many possible applications of critical path methods, or network planning techniques, that have come about since the development of PERT and CPM in the late fifties. Projects may involve routine procedures that are performed repetitively, such as the monthly closing of accounting books. In this case, network planning techniques are useful for detailed analysis and optimization of the operating plan. Usually, however, network planning techniques are applied to one-time efforts. Although similar work may have been done previously, it is not being repeated in the identical manner on a production basis. Consequently, in order to accomplish the project tasks efficiently, the project manager must plan and schedule largely on the basis of his experience with similar projects, applying his judgment to the particular conditions of the project at hand. During the course of the project, he must continually replan and reschedule because of unexpected progress, delays, or technical conditions.

Until just a few years ago, there was no generally accepted formal procedure to aid in the management of projects. Each manager had his own scheme, which often involved limited use of bar charts—a useful tool in production management but inadequate for the complex interrelationships associated with contemporary project management. The development of network based planning methods in 1957–1958 provided the basis for a more formal and general approach toward a discipline of project management. Critical path methods involve both a graphical portrayal of the interrelationships among the elements of a project, and an arithmetic procedure which identifies the relative importance of each element in the over-all schedule. Since their development, critical path methods have been applied with notable success to research and development programs, all types of construction work, equipment maintenance and installation, introduction of new products or services, or changeovers to new models, and even the production of motion pictures, conduct of political campaigns, and complex surgery. According to our definition, all of these activities are classed as projects.

In all of these projects one is, to a greater or lesser degree, concerned with developing an optimal (or at least a workable) plan of the activities that make up the project, including a specification of their interrelationships. Also, one is interested in scheduling these activities in an acceptable time span, and finally with "controlling" the conduct of the scheduled work. We might quote the old cliché that one should first plan the work and then work the plan.

With respect to planning and scheduling, one must consider the manpower and facilities required to carry out the program as it progresses in time. The aim is to plan the conduct of the program so that the cost and time required to complete the project are properly balanced, and excessive demands of key resources are avoided. With respect to the control function, one is concerned with monitoring the expenditure of time and money in carrying out the scheduled program, as well as the resulting "product" quality or performance. For the most part, critical path methods have concentrated on the time parameter and to a somewhat lesser extent on the cost parameter. The surface is just being scratched on the performance parameter; this is a much more difficult problem, since a technical judgment is required to assess performance.

DEVELOPMENT OF THE NETWORK PLAN CONCEPT

The heart of network based planning methods is a graphical portrayal of the plan for carrying out the program; such a graph shows the precedence relationships, i.e., the dependencies of the program's activities leading to the end objective. We call this graph a network. The network concept has developed in an evolutionary way over many years; it has been used for at least twenty-five years to display the courses and their prerequisites, which make up a particular college curriculum. In 1956, Flagle[1]† wrote a paper (published in 1961) on probability based tolerances in forecasting and planning, which was, in a sense, a forerunner to the development of PERT. However, it was not until 1957–1958 that project networking was formally defined concurrently by two research teams, one developing PERT (Program Evaluation and Review Technique)[2, 3] and the other CPM (Critical Path Method)[4, 5]. The approach taken in this text is that there is a "network based (critical path) planning methodology" which has emerged from CPM and PERT. This methodology is common to these two species, PERT and CPM, and to the other techniques that have since been developed that differ from these parents only in nonessentials. The features generally associated with CPM and PERT are treated as topics in themselves, to be applied as appropriate to the project circumstances, regardless of whether "CPM," "PERT," "Network Analysis," or another term is used to describe the general approach.

The development of PERT began when the Navy was faced with the challenge of producing the Polaris missile system in record time in 1958. Several studies[6, 7] indicated that there was a great deal to be desired with regard to the time and cost performance of such projects conducted during the 1950's. These studies of major military development contracts indicated that actual costs were, on the average, two to three times the earliest estimated costs, and the project durations averaged 40 to 50 percent greater than the earliest estimates. Similar studies of commercial projects indicated average cost and time overruns were 70 and 40 percent respectively. While many people feel that original estimates must be optimistic in order to obtain contracts, a more important reason for these failures was the lack of adequate project management planning and control techniques for large complex projects.

Admiral W. F. Raborn recognized that something better was needed

---

† Numbers refer to references given at the end of each chapter.

in the form of an integrated planning and control system for the Polaris Weapons System program. To face this challenge, a research team was assembled consisting of representatives of Lockheed Aircraft Corporation (prime contractor of Polaris), the Navy Special Projects Office, and the consulting firm of Booz, Allen and Hamilton. This research project was designated as PERT, or Program Evaluation Research Task. By the time of the first internal project report, PERT had become Project Evaluation and Review Technique. This research team evolved the PERT system from a consideration of technique such as Line-of-Balance,[8] Gantt charts, and milestone reporting systems.

Time was of the essence in the Polaris program, so the research team concentrated on planning and controlling this element of the program. As a result, one of the principal features of PERT is a statistical treatment of the uncertainty in activity performance time; it includes an estimate of the probability of meeting specified scheduled dates at various stages or milestones in the project. PERT also emphasizes the control phase of project management by various forms of periodic project status reports. The work of the original PERT research team has been extended into the areas of planning and controlling costs,[9] and to a lesser degree, into the areas of the performance or quality of the product.[10, 11]

CPM (Critical Path Method) grew out of a joint effort initiated in 1957 by the duPont Company and Remington Rand Univac. The objective of the CPM research team was to determine how best to reduce the time required to perform routine plant overhaul, maintenance, and construction work. In essence, they were interested in determining the optimum trade-off of time (project duration) and total project cost. This objective amounts to the determination of the duration of a project which minimizes the sum of the direct and indirect costs, where, for example, direct costs include labor and materials, while indirect costs include the usual items, such as supervision, as well as "cost" of production time lost due to plant downtime.

The activities comprising this type of project are characteristically subject to a relatively small amount of variation compared to the activities of the Polaris program. Hence, unlike PERT, CPM treats activity performance times in a deterministic manner and has as its main feature the ability to arrive at a project schedule which minimizes total project costs.

The pioneering PERT and CPM groups did not know of each other's existence until early 1959, when the momentum of each effort was too great to influence the other. However, the underlying basis of both CPM and PERT is the project network diagram.

The network diagram is essentially an outgrowth of the bar chart which was developed by Gantt in the context of a World War I military requirement. The bar chart, which is primarily designed to control the time element of a program, is depicted in Figure 1-1a. Here, the bar chart lists the major activities comprising a hypothetical project, their scheduled start and finish times, and their current status.

The bar chart, as well as other approaches which evolved from it, such as line-of-balance and milestone methods,* were not too successful on one-time-through projects, particularly those with a high engineering content. The important features of a project network that are designed to correct the deficiencies of the bar chart are that (1) the dependencies of the activities upon each other are noted explicitly, and (2) more detailed definition of activities is made. These two points will be discussed further below.

It is actually because of the first point that the network concept really became necessary. For modest-sized programs one can incorporate the dependencies on a bar chart implicitly. However, because of the enormous size of many present-day programs that contain thousands of significant activities, taking place in widely dispersed locations, it is no longer possible to treat interdependencies implicitly. This is also true for relatively small projects subjected to very detailed planning.

In the project network shown in Figure 1-1b, the lines denote activities which usually require time, manpower, and facilities to complete. Each activity originates and terminates in a unique pair of nodes called *events;* time flows from the tail to the head of each arrow. Events denote a point in time; their occurrence signifies the completion of all activities terminating in the event in question. For example, the occurrence of event 7 signals the completion of activities 3-7 (3→7) and 6-7. Dashed line arrows, called *dummies,* show precedence relationships only; they require no time, manpower, etc., to perform. Such a relationship is shown by activity 7-5. This networking scheme is the most widely used today; it is called activity-on-arrows, or merely an arrow diagram. Another way of networking is to reverse the role of the arrow and the node; the result is called activity-on-nodes, or a node diagram. The arrow diagram networking scheme will be used exclusively in Part I of this text because it is the most widely used method; however, the node method will also be illustrated in Part II of this text because of its potentialities.

Figure 1-1c shows the same project network drawn to scale on a

---

* See reference 11, pages 17 and 24.

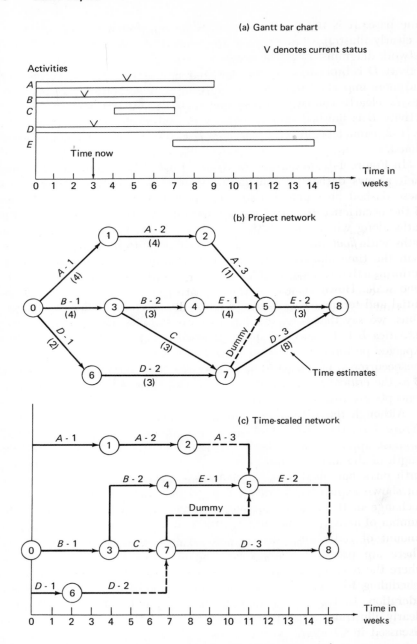

Figure 1-1   Comparison of Gantt Chart, Project Network and Time-Scaled Network.

time base. It is not too unlike the original Gantt bar chart; however, it clearly illustrates the two major differences in the bar chart and the network diagrams. First, the network shows greater detail. For example, activity D is broken down into activities D-1, D-2, and D-3. The second and more important difference is that the interdependency of the activities is clearly shown. For example, activity E can start as soon as activity B is finished; however, the last portion of activity E, denoted by E-2, cannot begin until activities E-1, A-3, C, and D-2 are all completed.

In Figure 1-1c, activities A-3, D-2, and E-2 have the last portion of their arrows dashed. Based on the estimated activity duration times, these dashed lines denote that these activities can be completed prior to the occurrence time of their succeeding events. For this reason, the paths along which these activities lie are referred to as *slack paths,* or paths with *float time;* that.is, these paths require less time to perform than the time allowed for them. For example, the estimated time to perform activities D-1 and D-2 is $2 + 3 = 5$ weeks, as indicated on the time scale. However, the time interval between the occurrence of the initial and terminal events of this path, i.e., events 0 and 7, is 7 weeks. Thus, we say that this path has two weeks of "slack" or "float." Now activities B-1, C, and D-3 have no dashed portions; the sum of their expected performance times, 15, is the same as the time interval between the occurrence of events 0 and 8. For this reason, this path is referred to as the *critical path;* it is the longest path through the network. These concepts are treated in detail in Chapter 4.

Although the network may be drawn to a time scale, as shown in Figure 1-1c, the nature of the network concept precludes this luxury in most applications, at least in the initial planning stage; thus, the length of the arrow is unimportant. You can "slide" activities back and forth on a bar chart with ease, because the dependency relations are not shown explicitly. However, if a network is drawn on a time scale, a change in the schedule of one activity will usually displace a large number of activities following it, and, hence, may require a considerable amount of redrawing each time the network is revised or updated. There are practical applications for time-scaled networks, especially where the network is closely related to production planning and where scheduling to avoid overloading of labor or facilities is a prime consideration. In these cases the time-scaling effort is worthwhile, for it clearly illustrates conflicting requirements. These applications will be discussed in Chapter 8 of this text, which deals with the allocation of resources.

VARIABLE VERSUS DETERMINISTIC ACTIVITY PERFORMANCE TIMES

To aid in determining the role of CPM or PERT in a particular project, it is useful to recognize that there are basically two different types of project activities and, unfortunately, all shades in between. If we view the actual time to perform an activity as a random variable, then we have, on the one hand, activities which perhaps have never been performed before, and which contain a considerable number of chance elements. They are called *variable activities,* and are characterized by a relatively large variance in their actual performance time. Examples of this type of activity are the design or development of new hardware items, the excavation of unsampled soil, or the performance of outdoor construction work during heavy rainfall seasons. The variable type of activity is shown on the left in Figure 1-2. On the other hand, we have a *deterministic* type of activity, i.e., one whose mean value is accurately known and whose variance in performance time is negligible, as shown on the right in Figure 1-2. Examples of deterministic activities are well-established maintenance operations and construction work.

Programs comprised primarily of variable activities (such as the development of the Polaris weapon system) may employ the PERT version of critical path methods. PERT emphasizes the control of the time element of program performance and treats explicitly the uncertainty in the performance times of the activities. The PERT system is based on three time estimates of the performance time for each activity: an optimistic (minimum) time, a most likely (modal) time, and a pessimistic (maximum) time. This system is treated in Chapter 11 which gives, among other things, the probability of meeting given

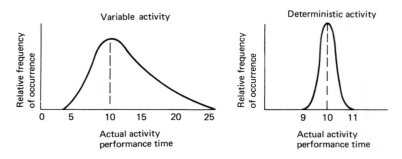

Figure 1-2 Variable versus deterministic activity performance times.

scheduled dates *without having to expedite the project activities.* On the other hand, programs comprised primarily of deterministic activities (such as construction or maintenance projects) utilize CPM, which omits the statistical considerations and is based on a single estimate of the average time required to perform the activity in question. Many projects, as will be pointed out below, can profitably employ facets of both methods.

The above distinctions between CPM and PERT are historically correct, and will be used repeatedly in this text. However, during the past few years a merging of these two techniques has taken place, and the result is, in most cases, referred to as a PERT-type system. In fact, some government agencies are apparently attempting to encompass all networking techniques with the title PERT. As an example, the National Aeronautics and Space Administration (NASA) in describing their NASA-PERT system, states that it has dropped the probability of meeting schedules in the output information; NASA-PERT permits the use of the single activity time estimate method characteristic of CPM.

ALLOCATION OF RESOURCES

All critical path methods emphasize the development of a workable plan and schedule for accomplishing the tasks making up the project. The network plan clearly displays the project activities and their order of performance. By the addition of time estimates for the performance of each activity, the project activities can be calendar oriented. Calendar orientation provides a convenient means of checking the schedule with respect to the utilization of key personnel or other resources. An example of such a check is shown in Figure 1-3. Thus, if the peak demand for

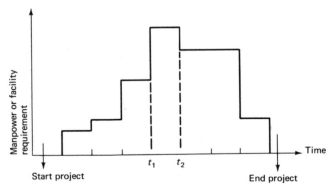

Figure 1-3 Manpower or facility feasibility requirement in time.

this particular resource (which occurs in the time interval, t_1 to t_2) is judged excessive, critical path methods provide a convenient means of replanning and scheduling the project to eliminate this peak. First, routine computations indicate where certain activity schedules can be moved forward or backward in time without affecting the completion time of the project. Then, critical path methods make it possible to simulate, in a very simple manner, the effects of various schedule shifts, or even changes in project plan, and in this manner to determine an acceptable way of eliminating the undesirable peak demand on the key resource.

TIME-COST TRADE-OFFS

The costs associated with a project can, for certain purposes, be classed as either direct or indirect. The *direct costs* typically include the items of direct labor and materials, or if the work is being performed by an "outside" company, the direct costs are taken as the subcontract price. The *indirect costs* may include, in addition to supervision and other customary overhead costs, items such as the interest charges on the cumulative project investment, and penalty (or bonus) costs for completing the project after (or before) a specified date. The time-cost trade-off problem is directed to the task of determining a schedule of project activities which considers explicitly the indirect as well as the direct costs, and in some cases attempts to minimize their sum.

Many industries (notably the construction industry) ordinarily assume that the best performance time for a specific activity is the one that minimizes the total direct costs of performing the activity. This time is usually longer than the minimum time required to carry out the activity because the utilization of overtime labor or more expensive types of equipment and materials is ruled out by the requirement of minimum direct costs. One reason why this somewhat conservative policy exists today is undoubtedly the lack of a planning and scheduling procedure whereby time-cost trade-offs can be accurately assessed. There are good reasons why this practice of choosing activity performance times which minimize direct costs is not optimal and why it should gradually disappear:

(1) During the conduct of the project in question, be it the construction of a highway or the development of a new product, resources are tied up in an unusable form. From the standpoint of the time value of money, this represents an expense which increases rapidly as the time required to complete a project increases. There is no

sound reason why the expense of late completion, or the "profit" from early completion should not be shared explicitly between the contractor and contractee.

(2) Since indirect (overhead) costs associated with a project increase with time, usually linearly, it is advantageous from this standpoint for the company performing the work to reduce the time required to complete the project.

(3) With the advent of critical path methods, time-cost trade-offs can be determined with sufficient accuracy for use in original bid proposals (wherein alternate completion times and costs may be given) or in the preparation of contract modifications.

By the utilization of the time-cost trade-off feature of critical path methods, an accurate relationship can be developed between direct project costs and the time required to complete the project. Then a total cost curve can be obtained by adding the indirect costs associated with the project. Ordinarily, the total cost curve will reach a minimum at some time short of the usual one based upon performing all activities so that direct costs alone are a minimum. This relationship is depicted in Figure 9-6 and is discussed in detail in Chapter 9. The use of the time-cost concepts in the preparation of bids is treated briefly in Chapter 12.

COST CONTROL

In addition to its value as a means of planning a project to optimize the time-cost relationship, the critical path network provides a powerful new vehicle for the control of costs throughout the course of the project. Most cost accounting systems in industry are functionally oriented, providing cost data by cost centers within the company organization rather than by project. By the utilization of the project network as a basis for project accounting, expenditures may be coded to apply to the activities, or groups of activities within a project, thus enabling management to monitor the costs as well as the schedule progress of the work.

Although the theory of network cost control is relatively simple, it is only just beginning to be employed as a practical supplement to basic critical path technology. The primary reasons for this late acceptance in practice have been the necessary involvement in established cost accounting procedures and the fact that computer programs developed for one company are not likely to be available or applicable to other companies. Thus, each organization interested in network cost control has been faced with the inconvenience and expense of developing new accounting procedures and computer programs. This situation is changing rapidly. Several large agencies of the U. S. Government[9] have begun to require the use of cost control supplements to basic CPM and PERT requirements, and generalized computer programs are being developed. Also, as the installation of computers expands further in industry, changes in accounting procedures are becoming more frequent. This expansion is becoming a necessity in many functionally organized companies that are experiencing problems of coordinating project activities because of rapid expansion of the volume of their work. Such firms are currently expending a great deal of effort to develop network based management information systems to alleviate this problem.

An introduction to the concepts and practical problems of network cost control is presented in Chapter 10. A discussion of the available computer programs for all types of network analysis is contained in Chapter 5.

SUMMARY OF NETWORK BASED PLANNING AND CONTROL PROCEDURES

The application of network based planning is a dynamic procedure. The end product, in the form of an acceptable project plan, is used as the basis of a closed loop feedback control system as shown in Figure 1-4. A summary of the steps involved in applying these critical path methods is given below along with references to Figure 1-4 and to appropriate chapters in the text.

STEP 1

Project Planning   The activities making up the project are defined, and their technological dependencies upon one another are shown explicitly in the form of a network diagram. This step is the subject of Chapter 2, Developing the Network, and is shown in box (1) of Figure 1-4. (Useful variations of the basic arrow diagram networking procedure are taken up in Chapter 6, Other Networking Schemes.) This is the most important step in the entire PERT/CPM procedure. The disciplined approach of expressing a plan for carrying out a project in the form of a network accounts for the majority of the benefits to be derived from PERT/CPM. It should also be added that if useful results are not obtained from these methods, it is almost always because of inadequately prepared networks.

STEP 2

Time and Resource Estimation   Estimates of the time required to perform each of the network activities are made; these estimates are based

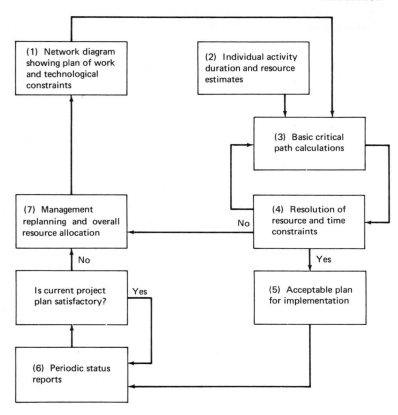

Figure 1-4   Dynamic network-based planning and control procedure.

upon assumed manpower and equipment availability and other assumptions that may have been made in planning the project in Step 1. This step is shown in box (2) of Figure 1-4. Single-time estimation is taken up in Chapter 3. The three-time estimation method associated with PERT is treated in Chapter 11.

STEP 3
Basic Scheduling   The basic scheduling computations give the earliest and latest allowable start and finish times for each activity, and as a byproduct, they identify the critical path through the network, and indicate the amount of slack or float time associated with the non-critical paths. This step, shown in box (3) of Figure 1-4, is taken up in Chapter 4. The problem of carrying out these computations in tabular form and, in particular, the problem of programming computers to carry out these basic computations is treated in Chapter 7.

STEP 4

Time-cost Trade-offs If the scheduled time to complete the project as determined in Step 3 is satisfactory, the project planning and scheduling moves on to a consideration of resource constraints in Step 5. However, if one is interested in determining the cost of reducing the project completion time, then time-cost trade-offs of activity performance times must be considered for those activities on the critical and near critical paths. This step, shown in box (4) of Figure 1-4, is taken up in Chapter 9.

STEP 5

Resource Allocation The feasibility of each schedule must be checked with respect to manpower and equipment requirements, which have not been explicitly considered in Step 3. This step, shown in box (4) of Figure 1-4, is taken up in Chapter 8. Establishing complete feasibility of a specific schedule requires frequent repetition of the basic scheduling computations, as shown by the recycle path from box (4) to box (3). It may also require replanning and overall adjustment of resources, as shown by the path from box (4) to box (7). Hence, establishing an acceptable project plan for implementation may require the performance of a number of cycles of Steps 3 and 4, and possibly Steps 1 and 2 as well.

STEP 6

Project Control When the network plan and schedule have been developed to a satisfactory extent, they are prepared in final form for use in the field. The project is controlled by checking off progress against the schedule, as indicated in box (6), and by assigning and scheduling manpower and equipment, and analyzing the effects of delays. Whenever major changes are required in the schedule, as shown in box (7), the network is revised accordingly and a new schedule is computed. The subject of time control is taken up in Chapter 4, and cost control in Chapter 10.

The basic procedures incorporated in Steps 1 through 6 can be performed, at least to some extent, by hand. Such methods will be presented in this text because they are useful in their own right, and also because they are an excellent means of introducing the more complex procedures which require the use of computers. It is particularly important that one be able to perform, by hand, the basic critical path calculations indicated in box (3) of Figure 1-4, since this is the first step in the evaluation of a proposed network plan for carrying out a project. A very simple method of hand calculation will be presented in Chapter 4. Hand methods for the resolution of relatively simple time and resource constraints will also be presented; they will then lead into more complex procedures for which computers are a necessity. Hand and computer methods of preparing periodic status reports will also be presented. A description of available computer programs is given in Chapter 5.

USES OF CRITICAL PATH METHODS

Since the successful application of PERT in the Polaris program, and the initial success of CPM in the chemical and construction industries, the use and further development of critical path methods has grown at a rapid rate. The applications of these techniques now cover a wide spectrum of project types.[12, 13]

Research and development programs range from pure research, applied research, development, to design and production engineering. While PERT is most useful in the middle of this spectrum, variations of it are now being used in the production end of this spectrum. PERT is not particularly useful in pure research, and in fact some say it should be avoided here because it may stifle ingenuity and imagination, which are the keystones of success in pure research.

Maintenance and shutdown procedures, an area in which CPM was initially developed, continues to be a most productive area of application of critical path methods. Construction type projects continue to be the largest individual area in which these methods are applied. It is extremely useful in this field of application to be able to evaluate alternate project plans and resource assumptions on paper rather than in mortar and bricks.

More recent applications of critical path methods include the development and marketing of new products of all types, including such examples as new automobile models, food products, computer programs, Broadway plays, and complex surgical operations.

In addition to an increase in the variety of applications of critical path methods, they are being extended to answer questions of increasing sophistication. The important problem of resource constraints has been successfully expanded to include multiple resource types associated with multiple projects. Cost control, project bidding, and incentive contracting are also areas where significant developments are taking place.

ADVANTAGES OF CRITICAL PATH METHODS

It is fitting to close this chapter with an enumeration of the advantages that one might expect from the use of critical path methods in the planning and controlling of projects.

(1) *Planning* Critical path methods first require the establishment of project objectives and specifications, and then provide a realistic and disciplined basis for determining how to attain these objectives, considering pertinent time and resource constraints. It reduces the risk of overlooking tasks necessary to complete a project, and also it provides a realistic way of carrying out more long-range and detailed planning of projects, including their coordination at all levels of management.

(2) *Communication* Critical path methods provide a clear, concise, and unambiguous way of documenting and communicating project plans, schedules, and time and cost performance.

(3) *Psychological* Critical path methods, if properly developed and applied, can encourage a team feeling. It is also very useful in establishing interim schedule objectives that are most meaningful to operating personnel, and in the delineation of responsibilities to achieve these scheduled objectives.

(4) *Control* Critical path methods facilitate the application of the principle of management by exception by identifying the most critical elements in the plan, focusing management attention on the 10 to 20 per cent of the project activities that are most constraining on the schedule. It continually defines new schedules, and illustrates the effects of technical and procedural changes on the overall schedule.

(5) *Training* Critical path methods are useful in training new project managers, and in the indoctrination of other personnel that may be connected with a project from time to time.

REFERENCES

1. Flagle, C. D., "Probability Based Tolerances in Forecasting and Planning," *The Journal of Industrial Engineering*, Vol. 12, No. 2, March-April (1961) pp. 97-101.
2. "PERT Summary Report, Phase 1," Special Projects Office, Bureau of Naval Weapons, Department of the Navy, Washington, D. C., July, 1958.
3. Malcolm, D. G., J. H. Roseboom, C. E. Clark, and W. Fazar, "Applications of a Technique for R and D Program Evalution," (PERT) *Operations Research*, Vol. 7, No. 5 (1959) pp. 646-669.
4. Walker, M. R., and Sayer, J. S., "Project Planning and Scheduling," Report 6959, E. I. duPont de Nemours and Co., Wilmington, Delaware, March 1959.
5. Kelley, J., "Critical Path Planning and Scheduling: Mathematical Basis," *Operations Research*, Vol. 9, No. 3, May-June (1961) pp. 296-321.
6. Marshall, A. W., and W. H. Meckling, "Predictability of the Costs, Time

and Success of Development," RAND Corp., Report P-1821, December, 1959.

7. Peck, M. J., and F. M. Scherer, "The Weapons Acquisition Process: An Economic Analysis," Division of Research, Graduate School of Business Administration, Harvard University, Cambridge, Mass., 1962

8. Turban, Efraim, "The Line of Balance—A Management by Exception Tool," *The Journal of Industrial Engineering*, Vol. 19, No. 9, September 1968, pp. 440-448.

9. *DOD and NASA Guide, PERT Cost Systems Design*, by the Office of the Secretary of Defense and the National Aeronautics and Space Administration, U. S. Government Printing Office, Washington, D. C., June, 1962, Catalog Number D1. 6/2:P94.

10. Malcolm, D. G., "Reliability Maturity Index (RMI)—An Extension of PERT into Reliability Management," *The Journal of Industrial Engineering*, Vol. 14, No. 1, January-February (1963) pp. 3–12.

11. Miller, R. W., *Schedule, Cost, and Profit Control with PERT*, McGraw-Hill Book Co., Inc., 1963, Chapter 6.

12. Bigelow, C. G., "Bibliography on Project Planning and Control by Network Analysis: 1959–1961," *Operations Research*, Vol. 10, No. 5, September-October (1962) pp. 728-731.

13. Wattel, H. L., "The Dissemination of New Business Techniques: Network Scheduling and Control (CPM/PERT), Hofstra University, Hempstead, New York, 1964.

EXERCISES

1. Discuss various applications of critical path methods. For example, suppose you are in charge of the preparation of a proposal for a large and involved project, or the coming church social, or the preparation of a new college curriculum, or the development of a new product and manufacturing facility. Would critical path methods be of assistance in these undertakings? If so, in what ways?

2. Can you think of any involved projects in which critical path methods would not be of any particular value? Give examples and discuss why.

2

DEVELOPMENT
OF THE
NETWORK

The first step in utilizing critical path methods is the identification of all the activities involved in the project and the graphical representation of these activities in a flow chart or network. This step is usually called the "planning phase," because the identification of the project activities and their interconnections requires a thorough analysis of the project, and many decisions are made regarding the resources to be used and the sequence of the various elements of the project.

In one sense the network is only a graphical representation or model of a project plan. The plan may have previously existed in some other form—in the minds of the project supervisors, in a narrative report, or in some form of bar chart. In practice, however, the preparation of a network usually influences the actual planning decisions and results in a plan that is more comprehensive, contains more detail, and is often different from the original thoughts about how the project should proceed. These changes derive from the discipline of the networking process, which requires a greater degree of analytical thinking about the project than does a narrative, a bar chart, or other types of project descriptions.

Thus, the construction of the network often becomes an aid to and an integral part of project planning, rather than an after-the-fact graphical exercise. Indeed, the planning phase has proven to be the most beneficial part of critical path applications. In developing a detailed and comprehensive project network, users often make significant improvements over their original ideas; they do a better job of early coordination with suppliers, engineers, managers, subcontractors, and all the other groups associated with the project; and they end up with a documented plan that has strong psychological effects on the future management of the project.

One psychological effect of network preparation is that it demonstrates to supervisors and other key personnel that the management is vitally concerned about the coordination and timeliness of all project activities, and that a means of more closely monitoring these factors has been drafted. Thus, an intangible but highly significant factor—the initial motivation of the project team—can be favorably influenced by the networking effort.

The planning phase is also the most time-consuming and difficult part of most critical path method applications. This is due primarily to the inherent analytical problems in any project planning effort. One may expect some difficulty in using the network format at first, but it is soon realized that the network discipline is more of an aid to thinking than it is a set of stringent rules for drawing a chart. Actually there are only about five rules to be followed in drawing networks, and these rules provide almost unlimited flexibility in describing project plans. The accuracy and usefulness of a network is dependent mainly upon intimate knowledge of the project itself, and upon the general qualities of judgment and skill of the planning personnel. Actual experience in drawing networks is important only in the first one or two applications.

This chapter is limited to those basic rules and procedures of network development which are required to prepare the first draft of a network. In Chapter 3 the addition of time estimates and the development of a final working draft are considered, including the problem of obtaining the most useful level of network detail.

There are several different graphical schemes used in drawing networks. This chapter and the rest of Part I will be based upon the most common scheme practiced among industrial users of critical path methods, that is, the *arrow* scheme. This is not necessarily the "best" scheme, however. Two of the other schemes, the *node* scheme and *precedence* diagramming, have much to recommend them, especially under certain conditions. Readers who are not limited by contract, the computer programs available, or other reasons to use a particular scheme are urged to study all of the schemes, which are presented in this chapter and in Chapter 6. The reader should then select the scheme best suited to his circumstances. As mentioned above, in the application of critical path methods the greatest effort and expense are associated with the preparation of the network. Selection of the most useful and economical networking scheme is, therefore, worthy of special attention.

PREPARATION FOR NETWORKING

Some experienced network users will say that all one needs to begin networking are a large piece of paper, several sharp pencils, and a large eraser. Actually, there is a bit more to it than that. As pointed out by Archibald and Villoria,[1] several general questions need to be raised and answered before detailed project planning should begin. The questions are:

(1) What are the project objectives?
(2) Who will be charged with the various responsibilities for accomplishing the project objectives?
(3) What organization of resources is available or required?
(4) What are the likely information requirements of the various levels of management to be involved in the project?

Of course, these questions are fundamental to project management and should not be passed over lightly. In some research and development projects, the development of new products, and in other cases, a discussion of the basic objectives of the project can reveal disagreements among the key persons involved. Similarly, open discussions of responsibilities and resources can bring to light erroneous assumptions or misunderstandings in these areas. Naturally, it is well to resolve these matters before proceeding with networking (and certainly before beginning the project).

An optional step in network preparation is the development of a list of work elements of the project. Such a list can be useful in discussing responsibilities and resources, as mentioned above, and it can serve as a reference for networking. Although experienced networkers usually forego the listing of activities, beginners with the technique will find that a list is helpful.

BASIC TERMS

Several of the most common terms in networking are defined and illustrated below. Terms associated with scheduling computations are explained in later chapters.

Definition:

An *activity* is any portion of a project which consumes time or resources and has a definable beginning and ending. Activities may involve labor, paper work, contractual negotiations, machinery operations, etc. Commonly used terms synonymous with "activity" are "task" and "job." Activities are graphically represented by arrows, usually with descriptions and time estimates written along the arrow (Figure 2-1).

Definition:

An arrow representing merely a dependency of one activity upon another is called a *dummy* activity. A dummy carries a zero time estimate. It is also called a "dependency arrow." Dummies are often represented by dashed-line arrows (Figure 2-2a) or solid arrows with zero time estimates (Figure 2-2b).

Definition:

The beginning and ending points of activities are called *events*. Theoretically, an event is an instantaneous point in time. Synonyms are "node" and "connector." If an event represents the joint completion

Figure 2-1

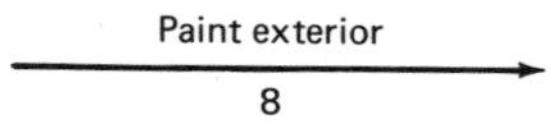

Paint exterior
8

Figure 2-2a

Figure 2-2b

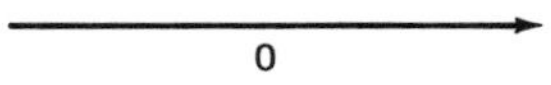

0

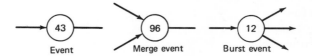

Figure 2-3

of more than one activity, it is called a "merge" event. If an event represents the joint initiation of more than one activity, it is called a "burst" event. An event is often represented graphically by a numbered circle (Figure 2-3), although any geometric figure will serve the purpose.

Definition:

A *network* is a graphical representation of a project plan, showing the interrelationships of the various activities. Networks are also called "arrow diagrams" (Figure 2-4). When the results of time estimates and computations have been added to a network, it may be used as a project schedule.

NETWORK RULES

The few rules of networking may be classified as those basic to all arrow networking systems, and as those imposed by the use of computers or tabular methods of critical path computation.

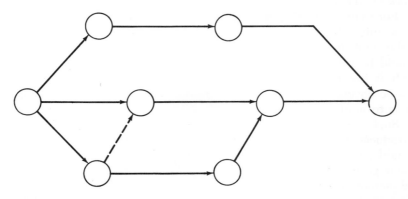

Figure 2-4

Basic Rules of Network Logic

RULE 1. Before an activity may begin, all activities preceding it must be completed.

RULE 2. Arrows imply logical precedence only. Neither the length of the arrow nor its "compass" direction on the drawing have any significance. (An exception to this rule is discussed under "Time-scaled Networks" below.)

Additional Rules Imposed By Some Computers or Tabular Methods

RULE 3. Event numbers must not be duplicated in a network.

RULE 4. Any two events may be directly connected by no more than one activity.

RULE 5. Networks may have only one initial event (with no predecessor) and only one terminal event (with no successor).

Rules 4 and 5 are not required by all computer programs for network analysis, as discussed in Chapter 6, nor are they required for methods of hand computation, as discussed in Chapter 4.

EMPHASIS ON LOGIC

At this point it should be noted that the construction of a network should be based on the logical or technical dependencies among the activities. That is, the activity "approve shop drawings" must be preceded by the activity "prepare shop drawings," because this is the logical and technically necessary sequence.

A common error in this regard is to introduce activities into the network on the basis of a sense of time.

For example, in the maintenance of a pipeline the activity "deactivate lines" might be placed after "procure pipe," because it is felt that that is the right time to deactivate the lines. Rather, the deactivation activity should be placed in the network in the proper technological sequence, such as just before "remove old lines." Then in the scheduling process (to be covered in Chapter 4) the best *time* to initiate the deactivation so as to minimize the down time on the pipe lines can be determined.

Such emphasis on strict logic is one of the principles of networking introduced by the originators of both CPM and PERT. It is a fundamental part of the networking discipline that causes planners to think about their project in a thorough, analytical manner. In this process old methods of performing similar projects are questioned or disregarded, clearing the way for new and perhaps better approaches.

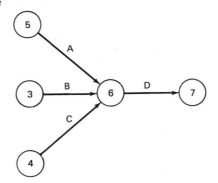

Figure 2-5

This principle is discussed further and an exception introduced under "Natural and Resource Dependencies" later in this chapter.

INTERPRETATION OF RULES

Rules 1 and 2 may be interpreted by means of the portion of a network shown in Figure 2-5. According to Rule 1, this diagram states that "before activity D can begin, activities A, B, and C must be completed." Note that this is not intended to imply that activities A, B, and C must be completed simultaneously.

Note also the definition of the events. Event 5 represents the "beginning of activity A." Event 6, however, means "the completion of activities A, B, and C, and the beginning of activity D." Because of the multiple meanings of events, discussion of networks in terms of activities is favored over event-oriented terms.

COMMON PITFALLS

The most common network error involves Rule 1. As an illustration, consider the diagram of activities A, B, C, and D shown above. Suppose that activity D depended on the completion of B and C and on the completion of the *first half* of A, completing the second half of A being independent of B, C, and D. To diagram this situation correctly, we must divide activity A into two activities and introduce a *dummy* activity, as shown in Figure 2-6. The dummy has been used here to correct a problem of *false dependency;* that is, activity D was only partially dependent on the activities preceding D. False dependencies represent the most subtle networking problems and must be guarded against constantly, especially at merge and burst points.

Another network condition that must be avoided is illustrated in Figure 2-7. Activities J, F, and K form a *loop,* which is an indication

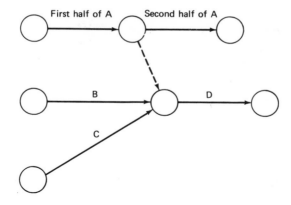

Figure 2-6

of faulty logic. The definition of one or more of the dependency relationships is not valid. Activity *J* cannot begin until *C* and *K* are completed. But *K* depends on *F*, which depends on *J*. Thus *J* could never get started because it depends on itself. Loops, which in practice may occur in a complex network through oversight, may be remedied by redefining the dependencies to relate them correctly.

SATISFYING COMPUTER RULES

Networking rules 3, 4, and 5 are related to the procedures for coding networks for computer analysis. Rule 3 involves another subtle problem that all computer programs have in understanding a network. Consider the diagram in Figure 2-8a. An attempt to process this situation on a computer would cause it to halt and print out an indication of loop, for the computer would read event 496 as a precedent to itself (Figure 2-8b). Therefore, in employing computers it is important to keep track of event numbers used and not used. (Some computer programs impose

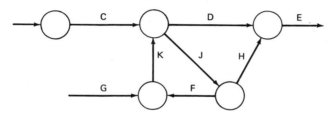

Figure 2-7

Figure 2-8a

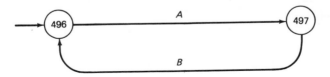

Figure 2-8b

the additional restriction that each event number must be larger than any predecessor number.)

Rule 4 is violated when the condition shown in Figure 2-9 occurs. Activities A and B may be called *duplicate activities,* since a computer (or tabular method of computation), using only event numbers for identification, may not be able to distinguish the two activities, as indicated below:

Network Description	Computer Code
Activity A	6-7
Activity B	6-7

One remedy calls for the introduction of a dummy and another event in series with either activity A or B (Figure 2-10). Now the computer can distinguish between the activities by their different codes.

Network Description	Computer Code
Activity A	6-7
Activity B	25-7
Dummy	6-25

Note that the above solution *does not change the logic* of the network. Nor would the logic be changed if the dummy had been placed at the other end of B, or at either end of A. If the reader feels that a change in logic has occurred, he should review the section Interpretation of Rules.

Another way to correct duplicate activities is to combine them (Figure 2-11). This solution is simple and effective, but it may destroy some

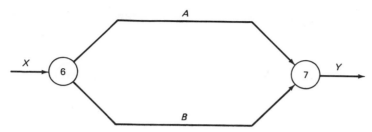

Figure 2-9

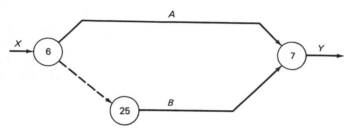

Figure 2-10

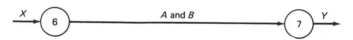

Figure 2-11

of the desired detail in the network; the question of detail is treated in Chapter 3.

Another special restriction for computer analysis is Rule 5. To accommodate this requirement it is common practice to bring all "loose ends" to a single initial and a single terminal event in each network, using dummies if necessary. For example, one may wish to network a current project that is already past the initial event. In this case the network would have a number of open ended, parallel paths at the "time now" point. These loose ends would be connected to a single initial event by means of dummies, as shown in Figure 2-12.

When methods employing hand computations on the network (or certain computer programs) are utilized, Rule 5 is not necessary.

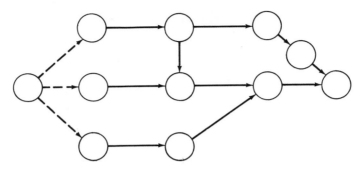

Figure 2-12

USING DUMMIES EFFICIENTLY

While the need for dummies in certain cases has been pointed out, it is preferable to avoid unnecessary dummies. For example, consider the diagram in Figure 2-13. Evidently, activity *D* depends on *C, B,* and *A*. But the dependency on *A* is clear without the dummy 2-4, which is *redundant.* Such dummies should be eliminated to avoid cluttering the network and to simplify computations. (Some computer services base their charges on the number of activities, including dummies.)

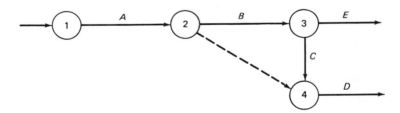

Figure 2-13

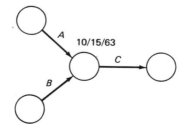

Figure 2-14

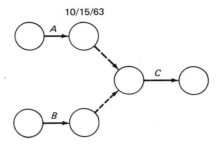

Figure 2-15

In other cases it may be necessary to introduce dummies for clarity. For example, suppose a particular event is considered a *milestone* in the project, a point that represents a major measure of progress in the project. This point may be assigned a scheduled or target date, which may be noted on the network as shown in Figure 2-14. However, the ambiguity of events, especially merge or burst events, can cause confusion about the target date notation. Does the date represent the scheduled completion of activity A, activity B, activities A and B, or the completion of activity C? In these cases it may be desirable to eliminate the possible misunderstanding by introducing dummies, as in figure 2-15. In this example it is now clear that the date refers to the completion of activity A. It is emphasized that this is only an illustration of the flexibility afforded by judicious use of dummies, and not a required practice at merge points or at milestones.

TIME-SCALED NETWORKS

The dashed line is sometimes employed for purposes other than to represent dummies. If a network is plotted on a time scale, dotted portions of

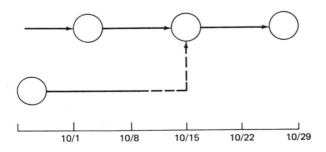

Figure 2-16

arrows may represent slack. The previous example might be shown as illustrated in Figure 2-16. The advantages and disadvantages of time-scaled networks are discussed in Chapter 12.

ACTIVITY DESCRIPTIONS

Thus far in this chapter, activities in the illustrative diagrams have been described by letter codes on the arrows. This has been done both for convenience and to emphasize the logic of the network representations. In practice it is much more common to print several descriptive words on the arrow. This avoids the need for cross-reference with a separate list of activity descriptors.

The descriptions themselves must be unambiguous; they must mean the same thing to the project manager, the field superintendent, the various subcontractors, and others expected to use the network. Descriptions should also be brief and, where possible, should make use of quantitative measures or reference points. Examples are shown in the following sample applications.

SAMPLE APPLICATIONS

A Machinery Installation

Consider a project involving the installation of a new machine and training the operator.[2] Assume that the training of the operator can begin as soon as he is hired and the machine is installed. The training is to start immediately after installation and is not to be delayed for inspection of the machine. The inspection is to be made after the installation is complete. One might attempt to network this project as shown in Figure 2-17.

However, this network says that the inspection cannot begin until the

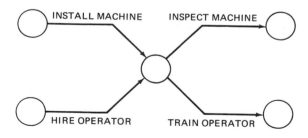

Figure 2-17

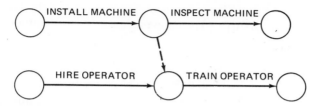

Figure 2-18

operator is hired, which is a false dependency. To correct this representation in the network a dummy is added, as shown in Figure 2-18.

A Market Survey

Consider now a project to prepare and conduct a market survey. Assume that the project will begin by planning the survey. After the plan is completed, data collection personnel may be hired, and the survey questionnaire may be designed. After the personnel have been hired and the questionnaire designed, the personnel may be trained in the use of the questionnaire. Once the questionnaire has been designed, the design staff can select the households to be surveyed.

Also, after the questionnaire has been designed it may be printed in volume for use in the survey. After the households have been selected, the personnel trained, and the questionnaires printed, the survey can begin. When the survey is complete the results may be analyzed. This project may be networked as shown in Figure 2-19. Note that dummy 4-3 is essential, whereas dummy 6-5 is necessary only if a computer is to be used.

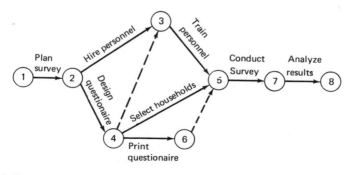

Figure 2-19

NATURAL AND RESOURCE DEPENDENCIES

The network in Figure 2-19 illustrates the fact that there are two types of activity dependency. Note that most of the activity dependencies are caused by the nature of the activities themselves; for example, personnel cannot be trained until they have been hired, and the questionnaires cannot be used in the survey until they have been printed. Such dependencies among activities may be called *natural,* and this is the most common type of dependency.

Also note, however, that the selection of households is dependent upon the design of the questionnaire, but only because one group of people is assigned to do both jobs (the "design staff" in the project description). This staffing limitation, and the implication that the design staff could not do both jobs simultaneously, causes the two activities to be drawn in series (dependent) rather than in parallel. A dependency of this type is not "natural," but is caused by the resource limitation. Thus, it may be called a *resource dependency.* The resources involved may be personnel, machinery, facilities, funds, or other types of resources.

Usually it is best to include in the first network draft all resource dependencies that are known and firmly established as ground rules for the project. These firm dependencies represent significant factors in the planning phase of a project and will often have major effects on the network and resulting schedule. However, if there is doubt about the number of resources available or how they should be allocated among the activities, then resource dependencies should be omitted in the first draft of the network, which would then be based only on natural logic. In these situations the techniques of resource allocation discussed in Chapter 8 should be employed.

Development of a New Product

The network in Figure 2-20 represents a plan for the development and marketing of a new product, in this case a new computer program.[3] Note that the dummy 5-7 is used to conform to Rule 4. The dummy 2-4, however, is used to show that activities 4-5 and 4-7 depend on activities 1-2 and 3-4, while activity 2-6 depends only on activity 1-2. Another dummy is shown between events 0 and 1, where the diagrammer has used an unconventional notation to indicate a "lead time"; this dummy is not technically necessary and could have been omitted by combining events 0 and 1. Note also that this network has two critical paths, denoted by the heavy activity lines.

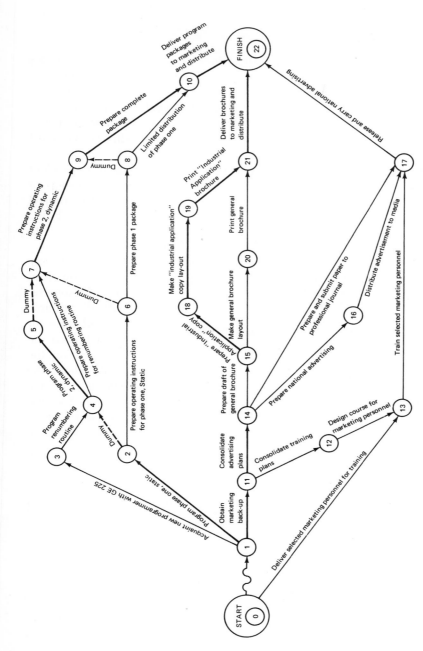

Figure 2-20 *Network for the development and marketing of a new computer program. (Courtesy Computer Department, General Electric Company)*

SUMMARY

This chapter has been concerned with the translation of the project plan into a series of interconnecting activities and events, composing a network model of the plan. A few rules were presented, some being required in order to maintain accuracy and consistency of network interpretation, and others being required by the nature of data-processing procedures. The rules presented relate to the arrow method of networking, although other networking methods are illustrated and compared in Chapter 6. The next chapter continues the development of the network, including the addition of activity time estimates and the attainment of the desired level of network detail.

REFERENCES

1. Archibald, Russell D., and Richard L. Villoria, *Network-Based Management Systems (PERT/CPM)*, John Wiley & Sons, New York, 1967.
2. Davis, J. Gordon, unpublished lecture notes for "Short Course in Project Management with CPM and PERT," Georgia Institute of Technology, Atlanta, 1967.
3. *GE 225 Application, Critical Path Method Program,* Bulletin CPB 198B, General Electric Computer Department, Phoenix, 1962.

EXERCISES

1. Review the machinery installation sample network in this chapter and assume that an activity consisting of "schedule inspector" must precede "inspect machine." Add the activity to the network without causing a false dependency.
2. In Figure 2-21, find at least five errors or unnecessary symbols. State which rule is broken in each case, and suggest how the error might be corrected.

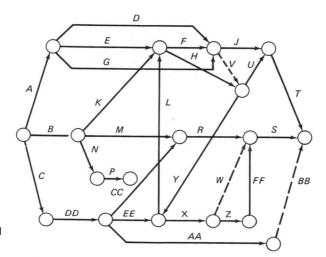

Figure 2-21

3. Given the activities and relationships listed below, draw an accurate network with no more than six dummies. Check your results by (1) numbering each event (do not repeat any numbers) and determining that no activity or dummy has the same pair of identification numbers, (2) making sure that there is only one initial and one terminal event.[2]

Activity	Predecessor
A	—
B	—
C	—
D	—
E	B, C, D
F	A, B, C, D
G	A, B, C, D
H	F, G, I
I	A, B, C, D
J	O, E, N
K	B, C, D
L	K
M	B, C, D
N	B, C, D
O	A, B, C, D

4. Using the list of activity dependencies given below, draw an accurate and economical (minimizing the use of dummies) network.

Activity	Depends on Activity		Activity	Depends on Activity
A	none		G	F
B	A		I	F
F	A		J	H
H	A		K	I and J
C	B		L	G, D, and E
D	B		M	K
E	C		N	L and M

5. Figure 2-22 shows a portion of a network for construction of a multi-story building. What do the dummies 29-33 and 34-38 represent? Can you find other dependencies of this type in the network?

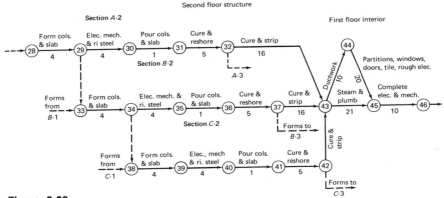

Figure 2-22

6. Assume that the following list of activity dependencies is correct. The diagram in Figure 2-23, however, does not represent these dependencies properly. Correct the diagram using only one dummy activity.

Activity	Depends on Activity
A	none
B	none
C	A
D	B and G
E	C and D
F	D
G	A
H	E

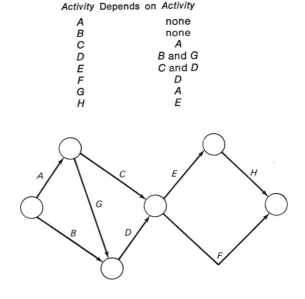

Figure 2-23

7. Draw a network of the following steam pipe maintenance project. The project begins by moving the required material and equipment to the site (5 hours). Then we may erect a scaffold and remove old pipe and valves (3 hours); while this is being done, we may fabricate the new pipe (2 hours). When the old pipe and valves are removed and the new pipe is fabricated, we can pace the new pipe (4 hours). However, the new valves can be placed (1 hour) as soon as the old line is removed. Finally, when everything is in place, we can weld and insulate the pipe (5 hours).

8. Consider as a project the servicing of a car at a filling station, including such normal activities as filling the gas tank, checking the oil, etc. List the activities you wish to include, then draw a network of the project. Assume that there are two men available to service the car and that the gas pump has an automatic shut-off valve.

In carrying out this exercise, remember that the network should reflect technological or physical constraints rather than arbitrary decisions on the order in which the activities are to be carried out. It is therefore suggested that you include time estimates for each activity and then redraw your original network on a time scale to show the actual order of scheduling each activity so as to minimize the time required for two men to service a car.

9. A common error in networking is to place the activities in the network according to the order in which you plan to carry them out, rather than in the order dictated by technological predecessor-successor relationships. Consider the following network.

The network was drawn this way because the planner felt that he would not start activity *D* until *B* and *C* were completed, and also that he would not start activity *E* until *D* was completed. If the true technological requirements on activity *D* are that only *A* must precede *D*, and no other activity depends on activity *D*, how should this network be drawn?

3
TIME ESTIMATES AND LEVEL OF DETAIL

By applying the networking rules presented in Chapter 2, one may develop the first draft of the project network. The basic logic of the plan should be established by the first draft. The next step is to add time estimates to each activity and refine the network as needed to reflect the desired level of detail.

In practice, the processes of time estimation and network refinement are closely interrelated and are usually accomplished at the same time. For as one begins to make time estimates, it is found that certain activities need to be redefined, condensed into fewer activities, or expanded into more, in order to represent the project accurately and at the desired level of detail.

As mentioned in Chapter 1, there are two methods

of applying time estimates: the single estimate method and the three estimate method. This chapter and the rest of Part I will treat only the single estimate method. The three estimates and their associated statistical treatment are considered in detail in Chapter 11.

TIME ESTIMATION IN THE NETWORK CONTEXT

Accuracy of Estimates

One of the most common first reactions of persons being introduced to critical path methods is that the whole procedure depends upon time estimates made by project personnel. Since these estimates are based upon judgment rather than any "scientific" procedure, it is argued, the resulting CPM or PERT schedules cannot be any better than schedules derived from bar charts or any other method.

While it is true that critical path methods depend upon human time estimates, as all project planning schemes must, there are some significant differences in how the estimates are obtained and in how they are used. To illustrate these differences, consider how estimates for project times and costs are usually derived. The process is similar in most types of industrial or construction projects, so let us take a familiar example, the construction of a house. A builder's estimate for a house may look something like this:

	Cost Estimate	Time Estimate
Clearing and grading	$ 450.00	—
Foundations	965.00	—
Framing	1,033.25	—
.	.	.
.	.	.
.	.	.
Total	$31,637.55	Approximately 3 to 4 Months

The total cost was developed from detailed, item-by-item estimates. The total time, however, was simply a gross estimate based on the builder's experience with similar projects. Careful attention to costs, of course, is the home builder's key to obtaining contracts and making a profit. He can afford to be less accurate about time estimates because the contracts do not normally have time limitations and the owner is not as concerned about the exact duration of the project.

But suppose the time were a critical factor, as it often is in industrial projects, and suppose the accuracy of the time estimate were made important to the builder's profit, through penalty clauses and other means. How could the builder develop a more accurate estimate of the project duration? We would expect him to break down the job into its time-consuming elements, to obtain good time estimates of each element, and to sequence the elements into a plan that would show which elements must be done in series, which in parallel, etc. In principle we are saying that greater accuracy in time estimation can be developed in a manner similar to the development of accurate cost estimates—through detailed, elemental analysis.

Critical path methods provide a disciplined vehicle for making detailed time estimates, for graphically representing the sequence of the project elements, and for computing the project duration. The facts that the network approach lends itself to a greater degree of detail than does a bar chart, and that the network shows the sequential relationships explicitly, are reasons for the belief that networks provide a greater degree of accuracy in the application of the knowledge of time estimators. The human knowledge is still the basic ingredient, but it is how the knowledge is applied that affects the accuracy of the results.

Who Does the Estimating?

Certainly another key factor in the accuracy of time estimates is who makes them. A general rule in this regard is that the most knowledgeable supervisory person should estimate each activity. This means, for example, that activities that are the primary responsibility of the electrical subcontractor should be estimated by the subcontractor's manager or supervisor most familiar with the job; activities of the research department should be estimated by the research supervisor responsible for and most familiar with the work; and so on. The objective in obtaining times estimates should be to get the most realistic estimates possible.

It is characteristic of CPM and PERT planning to call meetings of all supervisory personnel at the time-estimating stage and to consider each activity for which they are responsible. (It is also desirable for these personnel to participate in preparation of the first network draft. This is not always practical, though, if everyone involved is not familiar with network principles.) In addition to the psychological advantages mentioned in Chapter 2, the participation of the key members of the project team has major advantages. Whenever the subcontractors, suppliers, inspectors representing the customer, etc., meet to discuss a project, the discussion will lead to questions of priorities in certain phases of the work, potential interference of work crews, definitions of assignments of engineers, and many other details of planning that might not have been explored until problems arose during the project. These discussions often identify and resolve potential problems before the project begins, rather than tackling them as they actually occur, which can mean the corrective action will be expensive or perhaps impossible.

Here again the network merely serves as a cause for calling the meeting and as a detailed agenda. Yet these thorough planning sessions around a network result in what is probably the major benefit of critical path methods as practiced to date. This benefit is the project plan itself, in terms of its validity, its comprehensive scope, and its efficiency in the utilization of time and resources. This is not to say the reader should stop here or even at the end of Part I of this text. There is more to be gained, and many of the more experienced users are applying critical path methods and related procedures to advance the science of project management in a variety of ways. Historically, however, it appears that a majority of satisfied users during the first ten years of CPM and PERT have gotten their money's worth out of the initial network planning effort and have not followed through with the technique in the project control applications.

The question of bias or "padding" of time estimates is, naturally, related to who makes the estimates and their motivations. It does not necessarily follow that the most knowledgeable person is also the most objective. It is human nature to try to provide a time estimate that will be accepted as reasonable but will not likely cause embarrassment later. Thus, a certain amount of bias is to be expected in any procedure.

Nevertheless, it is generally felt by CPM users that the network approach tends to help reduce the bias to a manageable level. Again, the increased detail shown by a network plays a useful role. The smaller the work elements, the more difficult it is to hide a padded estimate. Indeed, a certain amount of professional pride is often noticeable in the estimator's attitude, which leads to a degree of optimism in his figures. Another factor favoring realistic estimates is the recognition that biased figures will tend to make the activities involved form the critical path, thus invoking concentrated attention of management and other parties engaged in the project on the group responsible for those activities.

Research on Time Estimates

Only a limited amount of research on the subject of network time estimation has been published. One of the more interesting studies, by Seelig and Rubin,[1] compared the results of 48 R & D projects, some of which were "PERTed" and some of which were not. The authors concluded that the use of PERT definitely did lead to improvement of schedule performance but had no noticeable effect on technical performance. Furthermore, they concluded that the improvement in schedule performance was primarily a result of *improvement in communication* among the project managers, which was brought about by the use of PERT.

ESTIMATION METHODS

When to Add Time Estimates

It is best to complete a rough draft of the total project network before any time estimates are added. This procedure is conducive to concentration on the *logic* of the activity relationships, which must be accurately established. When the draft appears to be complete, the time estimates should be added to each activity. This step will constitute a complete review of the network, and will usually result in a number of modifications based on the diagrammer's new perspective of the total project network.

As soon as the estimates are completed, a simple hand computation of the forward pass should be made (this results in the earliest activity start and finish times, as explained in the next chapter). This is an important step, for it may reveal errors or the need for further refinement before the preparation of the final working draft of the project network.

Conventional Assumptions

The time estimate to be made for each activity is called the *activity duration*. This term is employed to imply the elapsed time of the activity expressed in units such as working days, rather than a measure of effort expressed in units such as man-days. Units other than working days, such as hours or weeks, may be utilized, provided the unit chosen is used consistently throughout the network. Estimates of activity duration do not include uncontrollable contingencies such as fires, floods, strikes, or legal delays. Nor should safety factors be employed for such contingencies.

In estimating an activity's duration time, the activity should be considered independently of activities preceding or succeeding it. For example, one should not say that a particular activity will take longer than usual because the parts needed for the activity are expected to be delivered late. The delivery should be a separate activity, for which the time estimate should reflect the realistic delivery time.

It is also best to assume a normal level of manpower, equipment, or other resources for each activity. Except for known limitations on resources that cause some activities to be resource dependent (discussed in Chapter 2), do not attempt to account for possible conflicts between activities in parallel that may compete for the same resources. These conflicts will be dealt with later, after the scheduling computations have been made.

Accounting for the Weather

In construction projects the weather is one of the greatest sources of scheduling uncertainty. In a single-estimate system, there are two common approaches for taking the weather into account.

The first approach is to omit consideration of the weather when estimating the duration of each activity, and instead, estimate the total effect of weather on the project's duration. For example, suppose a project's duration is computed to be 200 working days. Consideration is now made of the seasons in which the outdoor work will be done, the seasonal temperatures and precipitation in the region, the type of soil, type of construction, and other weather-related factors. It may be estimated that five weeks would be lost because of bad weather. Thus the total project duration would be increased to about 225 working days. However, this approach is no different from the usual method of accounting for the weather in construction estimating. It does not take advantage of the detailed breakdown of activities afforded by the project network.

The second approach involves the consideration of weather effects in making each activity time estimate. In this approach each activity is evaluated as to its weather sensitivity—excavation work being sensitive to rain, concrete work sensitive to freezing, interior plumbing not weather sensitive, etc. Suppose an activity is estimated to require ten man-days, and two men will be placed on the task; the nominal time estimate will be five working days. The weather sensitivity of this activity, the season, and other weather factors may indicate that this activity's estimate duration should be increased about 20 per cent. Thus, the adjusted time estimate is six working days.

The advantage of this detailed approach to weather adjustments is that it applies the adjustments to particular portions of the network, which will result in a more accurate schedule for each activity with reference to calendar dates. A disadvantage is the need to add more notations to the activity descriptions. Activity descriptions, including time estimates, must be clearly understood by all persons expected to work with them. This means that both nominal and weather-adjusted estimates should be noted on each activity; in some cases, it is desirable

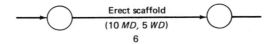

Figure 3-1

to add the man-day estimate as well. With these notations, an activity may appear as in Figure 3-1, where MD = man days, WD = working days. A legend on the network should explain the notations, including the fact that the number standing alone represents the weather-adjusted estimate in working days.

Accounting for Weekend Activities

The use of working days, which is a common time unit in construction projects, results in computations of *project duration* which assume that no activities proceed on weekends and holidays. However, this may be incorrect. For example, concrete may be cured and buildings may be dried out over nonworking days. In such cases, time estimates in working days tend to result in an overestimate of the project duration. When activities of this type are expected to take longer than 5 calendar days, the overestimate can be corrected to a certain extent by a suitable adjustment of the working-day estimate. For example, a curing requirement of 6 or 7 days can be estimated as 5 working days (assuming a five-day work week). A curing requirement of 5 days, however, may actually take 5 working days, although it would likely run over a weekend and thus consume only 3 or 4 working days; in such instances, the estimator should employ the project network to judge the likelihood of curing over the weekends and adjust the working day estimate accordingly.

All time estimates in a network must be based on the same number of working days per week. For activities that will deviate from this standard, adjustments must be made in the time estimates similar to the adjustment for curing activities mentioned above.

ACTIVITY REDEFINITION

As in networking, the proper application of time estimates depends primarily on judgment and experience with critical path techniques. Illustrated here are some of the common networking problems uncovered when one is attempting to make time estimates. In most cases the problems involve activity definitions and the question of detail. Alternative solutions to each problem are discussed.

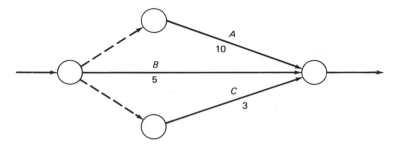

Figure 3-2a

Activities in Parallel

In the situation illustrated in Figure 3-2a several interrelated activities may begin and end at approximately the same nodes.

Suppose that upon supplying the time estimates, however, it is realized that A cannot begin until B has been underway for one day, and will not be completed until one day after B is completed. Furthermore, C can be accomplished at any time B is going on. A practical illustration of interrelationships of this type is the work prior to pouring a slab of concrete in a building, where the steel, mechanical, and electrical trades may begin and end their work at different times, but during much of the time they are all working in the same area. To network situations of this type accurately, we must redefine the activities, remembering that an activity is any portion of the work that may not begin until other portions are completed. Using this definition and the conditions prescribed above, we may correct Figure 3-2a as shown in Figure 3-2b.

Note the use of percentages to define activities in this illustration. This is occasionally a useful device, but frequent use of percentages is

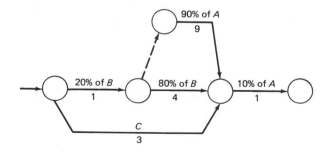

Figure 3-2b

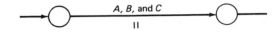

Figure 3-2c

not good practice. They often represent arbitrary definitions which can lead to misunderstandings by subcontractors and others involved in interpreting the network. Where possible, it is better to use physical measures, such as yards of concrete poured, the number of columns formed, the specific items assembled, etc.

Another approach to the correction of the activities-in-parallel problem would be to condense them into a single activity. However, the single activity should have a time estimate representing the total time for the completion of all three activities, a time estimate which is most accurately obtained from a detailed solution as given in Figure 3-2b. From this figure it is clear that the total time required for all three activities is 11 days. Thus the condensed activity may be represented as shown in Figure 3-2c.

Whether 3-2b or 3-2c is the "best" solution to the problem depends on the project, the network objectives, the areas of responsibility involved, and other factors which can be resolved only through the judgment of the project manager. This is essentially the problem of the level of detail, which will be discussed further. It should be noted here, however, that accuracy and detail are directly related, and even when less detail is desired in the final draft of the network, it is often useful in time estimations to sketch certain portions of the network in greater detail.

Activities in Series

Let us look at another problem of network accuracy, this one arising in a portion of a network in which the activities are drawn in series, as shown in Figure 3-3a. Suppose that upon inspecting this network it was realized that A and B did not require a total of 17 days. Actually part of B could begin at least 2 days before A was completed. (This is similar to the previous problem, since we are saying that A and B are partly concurrent.) An erroneous diagram of this type may be corrected in several ways. One way would be to split A at the point that B begins

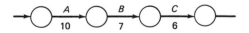

Figure 3-3a

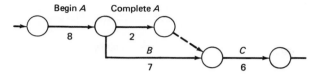

Figure 3-3b

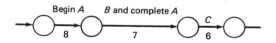

Figure 3-3c

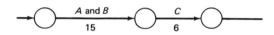

Figure 3-3d

(Figure 3-3b). Another way would be to simply absorb the completion of A in B, making sure that the activity descriptors were clear (Figure 3-3c). A third way, of course, would be to condense A and B (Figure 3-3d).

Note that in each of the alternative solutions illustrated in Figures 3-3b, 3-3c, and 3-3d, the total elapsed time for A and B is shown correctly as 15 days. Again, the choice of a solution depends on what one wishes to illustrate and control, and on such factors as the magnitude of the times involved and the feasibility of defining the activity segments clearly.

Practical Example

The problems discussed above arise repeatedly in practical efforts to draw accurate networks. Therefore, it is worthwhile to review the points made in terms of a practical example. Assume that the project is the digging and pouring of the footings for a large building. (For simplicity let "pour" include the placing of steel.) Assume that we have one digging crew and one pouring crew. It is estimated that the work will require three days of digging and three days of pouring. The network for this project might be drawn as shown in Figure 3-4a.

It will be obvious to readers familiar with construction that this diagram is unrealistic. The digging and pouring activities would not take six days altogether, for some of the time both activities will be concurrent. How, then, do we diagram the process accurately? A good approach is to subdivide the footings in some convenient manner. In this case we may divide them into sections the size of one day's digging. Call them Sections A, B, and C. Now the network may be drawn as shown in

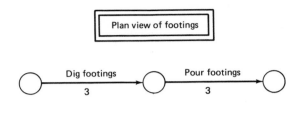

Figure 3-4a

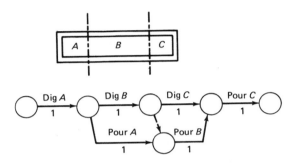

Figure 3-4b

Figure 3-4b, where we now see that the digging and pouring work should require only four working days.

NETWORK CONDENSATION

In the foregoing illustrations the concepts of condensing and expanding networks are introduced for the purposes of improving accuracy, eliminating excessive detail, and to achieve other objectives in the development of the detailed network. There are also occasions in which it is desired to produce a summary network for review by top management, which calls for the same condensation concept illustrated above, except that it is applied on a broad basis throughout the network, with the purpose of developing a general condensation and summarization of the project plan. In practice this often means that a network of several hundred activities must be reduced to one of a few dozen activities, without distortion of the logic, such that a summary picture of the project may be presented for review by top management, a customer, or other interested audiences. Consequently, some points related to condensation procedures are worthy of attention.

In general, a safe rule of condensation is that groups of activities independent of other activities may be condensed without distorting the

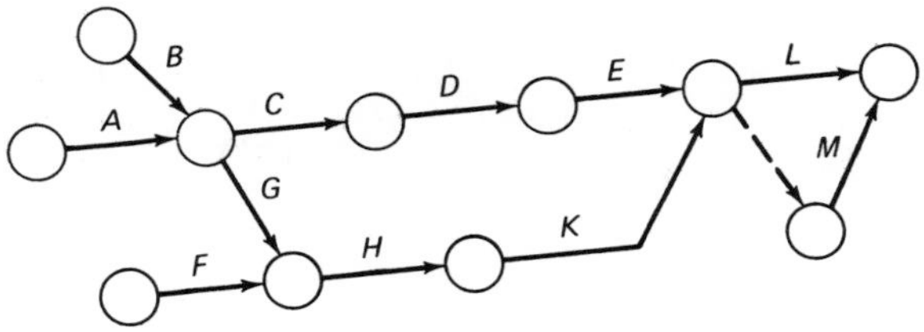

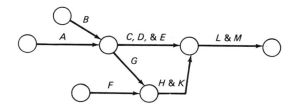

Figure 3-5a

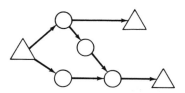

Figure 3-5b

network logic. For example, consider Figure 3-5a. In this network there are three independent activity groups that may be condensed, as shown in Figure 3-5b. Note that some activities in series have been combined, and in one case two activities in parallel, *L* and *M*, were combined. *But all the dependency relationships in the original diagram still hold.* This is the most important point in condensing networks, for it is very easy to introduce false dependencies.

A somewhat different approach to condensation is used by certain computer routines that perform this function. This approach calls for the designation of certain key events in the network which are not to be omitted in the condensation procedure. Then all direct and indirect restraints (groups of activities) between each pair of key events are reduced to a single restraint (activity). To illustrate, consider the network of seven activities in Figure 3-6a. The triangles denote the selected key events. Using the condensation process described above, this network would be reduced to the two activities shown in Figure 3-6b.[2]

Figure 3-6a

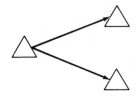

Figure 3-6b

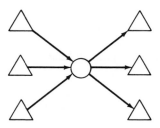

Figure 3-7a

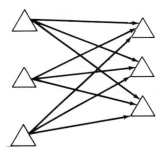

Figure 3-7b

This particular procedure is vulnerable to the occurrence of "pathological cases," in which the number of activities is not reduced or may even be increased. Using the procedure on the network in Figure 3-7a, for example, produced the network in Figure 3-7b. However, by selecting key events with this possibility in mind, one can avoid most pathological cases.[1]

THE LEVEL OF DETAIL

Thus far, comments on the problem of the level of detail in a network have been associated with questions of accuracy and economy of the presentation. There are many other factors involved in determining the most appropriate level. In considering any particular activity or group of activities with regard to expanding, condensing, or eliminating it, the diagrammer may ask himself several questions to guide his decision:

(1) Who will use the network, and what are their interests and span of control?

(2) Is it feasible to expand the activity into more detail?

(3) Are there separate skills, facilities, or areas of responsibility involved in the activity, which could be cause for more detail?

(4) Will the accuracy of the logic or the time estimates be affected by more or less detail?

Clearly, these questions are only guides to the subjective decision that must be made in each case. Generally, after working with one or two networks, a person will develop a sense for the appropriate level of detail.

That there are no firm rules that may be followed in determining the level of detail is illustrated by the following hypothetical case. The project is the construction of a house. If the network rules of Chapter 2 are followed, one could prepare a complete network, as shown in Figure 3-8a, or one could take a more detailed approach, as indicated in the portion of the network shown in Figure 3-8b.

These appear to be clear examples of too little and too much detail. But suppose the house is one of a hundred identical ones in a large housing project. Three activities per house would thus result in a network of 300 activities, plus other activities for roads, utilities, etc. Such a broad network may be very useful in analyzing the over-all length of the project, the most desirable sequence of construction, and other problems of general planning. Furthermore, since the house construction in this case is a matter of mass production, it would be worthwhile to work

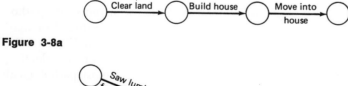

Figure 3-8a

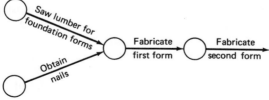

Figure 3-8b

out the construction schedule for one house in considerable detail, for any bottlenecks in the schedule for one house would cause repeated delays in all houses. Thus the detailed network treatment for a typical house might well be justified under these conditions. Under most other conditions the approaches illustrated above would indeed represent a useless extreme on the one hand and an expensive, perhaps impractical extreme on the other.

Cyclical Networks

In the house building case cited above, it was stated that both detailed and condensed networks may be useful if a number of identical houses are to be built. It may be generally stated that whenever a project involves a number of cycles of a group of activities, one should consider (1) developing a detailed network of the group, (2) condensing the detailed network into a summarized version, and (3) using the condensed network in the cycles that comprise the total project network. The purposes of the detailed network are to develop an efficient plan for the group of activities that will be repeated and to derive accurate time estimates for the condensed version. The purpose of the condensed version is network economy, since repetition of the detailed network would be costly in drafting time and would unnecessarily complicate and enlarge the total project network. Project types to which this principle would apply include multistory buildings, bridges, pilot production of a group of missiles, and a series of research experiments.

An example of the application of detailed and condensed networks in a multistory building is shown in Figure 3-9. Here the contractor worked out the detailed network in hours in order to balance the crews and minimize delays in the structural work on a floor. Then the condensed version was repeated for each floor of the building in the total project network.

NETWORK ORGANIZATION
The Pyramid Approach

In the large research and development programs of the aerospace industry, literally tens of thousands of activities and scores of networks may be involved. In order to make practical use of network systems of this magnitude, it is necessary to organize the networks by the various subsystems in the program, and to condense them into several different levels of detail. The different levels of detail are chosen to correspond to the levels of management of the program.

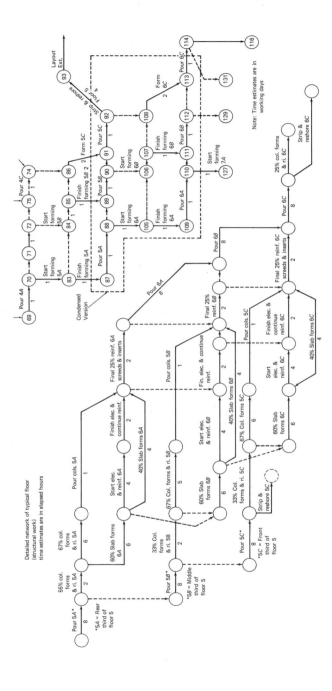

Figure 3-9 Portion of network for eleven-story building project showing detailed and condensed versions of structural subnetwork. *(Courtesy Floyd D. Traver & Co., Atlanta)*

55

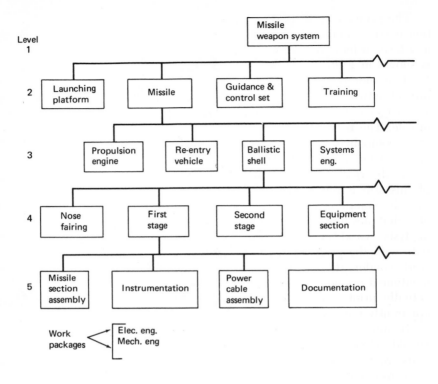

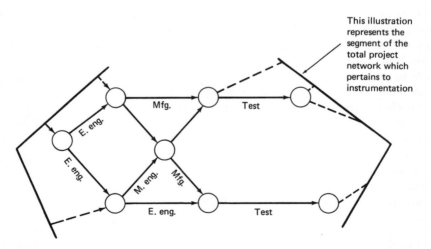

Figure 3-10 Example of a work breakdown structure. (*From the DOD-NASA Guide*[3])

The primary approach developed for this kind of network organization is based upon a description of the end product and its components in a level-of-indenture chart or "work breakdown structure." An example is shown in Figure 3-10. Each level of this chart identifies not only a level of indenture of the hardware but also a level of detail for the project networks. This particular chart, for example, implies that there is a management position at Level 4 responsible for the first stage of the missile. The networks reviewed at this level would include all the first stage components, shown in Level 5, but would not necessarily include all the detailed networks for each component.

As the reader may imagine, the monthly maintenance of networks of this size is not a part-time job. Many network analysts and large-capacity computer programs are required. In this environment a number of techniques have been developed and attempted in order to reduce the analysts' workload and enable the network procedures to keep up to date with the progress of the program. One such technique is the network condensation feature of certain computer programs, which was mentioned previously. Attempts have also been made to have computers actually print updated, time-scaled networks, although this effort has apparently not proven to have enough capacity to be practical.

In most industrial and construction applications, projects involving 50-500 activities are more typical. Elaborate schemes of network pyramids are not needed, but some principles of network organization should be considered. For instance, one level of condensation may be well worthwhile whenever it is necessary or desirable to communicate the progress of a large project to top management or to the customer. Suppose a machinery installation project contains 500 activities, and a company officer who is several levels above the plant superintendent desires to be kept informed of the progress. In such a case it may be worthwhile to prepare a time-scaled network of 50-75 activities and update it periodically as an aid to progress reports for the officer.

Another more common consideration should be the organization and legibility of the detailed network itself. A large, free-flowing network without clear groupings of related activities can be difficult to read and thereby somewhat self-defeating in its purpose. Usually with only a small amount of effort a network can be organized into subnetworks (or "subnets") which can be identified by large labels, border lines, or other means that improve the visibility of the subsystems in the project and their interrelations. An example of this kind of subnet identification is shown in Figure 3-11. Even in such a small and uncomplicated network the blocked areas and labels greatly aid the viewer in quickly comprehending the scope and content of the project. When blocking of this

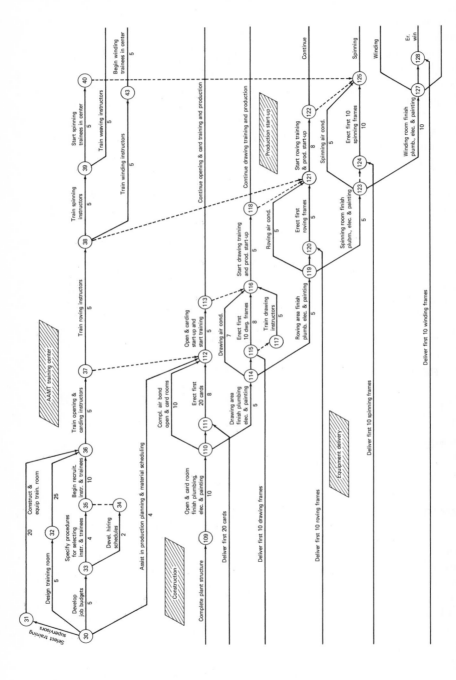

Figure 3-11 Portion of a network for a new plant start-up project showing labels for major sections of the project. *(Courtesy Textile Industries)*

58

type is well done it can eliminate the need for separate, condensed networks.

In most industrial and construction networks the best organizational approach is by physical subdivisions of the project—for example, by the floors of a building, the spans of a bridge, the components of an electronic product, etc. This organization provides a ready correspondence between the physical progress of the work and the graphic progress on the network.

Within each physical subdivision the network may be organized into further physical groups or into responsibility groups. The responsibility groups aid the supervisor of each group in the management of his activities. However, it is usually impractical to obtain complete organization by responsibility, due to the interrelationships of various responsibilities within each physical subdivision. As an illustration of this problem, consider the responsibility of the electrical subcontractor in Figure 3-9. The electrical activities, being less constraining on the schedule than the structural work, are subordinated in the network organization and appear scattered throughout the diagram. The supervisor of the electrical work does not see his activities all together in a neat group, but he does see them in their true dependency relationships with the other trades in the project.

CASE REFERENCE

Several interesting case studies of critical path applications have appeared in trade journals over the past four years. One of these is an experience of the Pure Oil Company,[5] which illustrates many of the techniques of network organization and judgment as to the appropriate level of detail. One of the networks of this project is shown in Figure 3-12 and discussed in Exercise 2.

SUMMARY

Although the preparation of the network is only the first phase in applications of critical path methods, many users have reported that the greatest benefits from the critical path concepts were derived from this phase alone. They felt that preparing the network caused them to think through the project in a more complete manner than ever before, forcing them to do a more thorough job of advanced planning. However, a great deal of useful information is included in the completed network, and the proper processing and utilization of this information as described in the following chapters can bring important additional benefits not only to

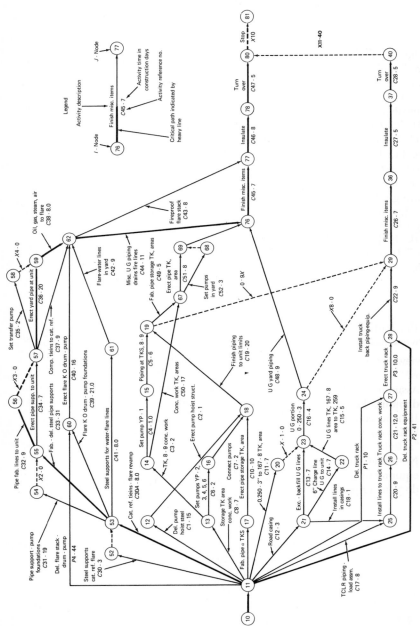

Figure 3-12 Example of a network for a piping installation project. *(Courtesy Heating, Piping, and Air Conditioning[5])*

the project manager but also to the subcontractors and all other groups engaged in the project effort.

REFERENCES

1. Seelig, W. D., and I. M. Rubin, "The Effects of PERT in R & D Organizations," published as a Working Paper of the Research Program on the Management of Science and Technology, No. 230-66, Alfred P. Sloan School of Industrial Management, Massachusetts Institute of Technology, December, 1966.
2. *IBM 7090 PERT COST Program*, IBM Program Application Bulletin H20-6297, International Business Machines Corporation, Data Processing Division, 112 East Post Road, White Plains, N. Y., 1962.
3. *DOD and NASA Guide, PERT Cost Systems Design,* Department of Defense and National Aeronautics and Space Administration, Government Printing Office, Washington 25, D. C., June, 1962.
4. Fry, James C., et al., "Managing Technological Change," *Textile Industries,* August, 1967.
5. Mark, E. J., "How Critical Path Method Controls Piping Installation Progress," *Heating, Piping, and Air Conditioning,* September (1963) pp. 121–126.

EXERCISES

1. In the footings project illustrated in Figures 3-4a and 3-4b, suppose the network had been drawn as shown in Figure 3-13. What error is involved?
2. Review the network shown in Figure 3-12 and answer the following questions.
   a. Assuming a computer was used to process this network, could any of the dummies be eliminated without distorting the logic of the network?
   b. If hand computations are used, which dummies could be eliminated?
   c. What do you suppose the diagrammer accomplishes by the use of the "activity reference number?"
   d. Note that the delivery of certain items, such as truck rack and truck rack equipment (activities 11-27, 11-27), are included in the network. Why do you suppose the diagrammer did not include the delivery of all items,

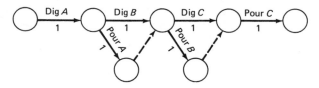

Figure 3-13

such as concrete, pumps YP-1 and YP-2 (activities 13-16, 14-15), and insulation (activities 36-37, 77-78), etc.?

  e. How could the visibility of this network be improved?

3. Refer to Figure 2-22. Explain how this network is organized.

4. Refer to Figure 2-20. Sketch blocks around portions of this network to identify the subsystems involved in the project.

5. The network in Figure 3-14 contains 17 activities. Condense it to 10 activities or less, without distorting the dependency relationships.

6. Condense the network in Figure 3-15 without distorting the logic.

4

BASIC SCHEDULING COMPUTATIONS

At this stage in the application of critical path methods, the project network plan has been completed and the mean performance times have been estimated for each activity. We now consider the questions of how long the project will take and when activities may be scheduled. Answers to these questions are inferred from the arrow diagram and the estimated durations of the individual activities. These estimates may be based on a single time value, as described in Chapter 3, which is basically the original CPM procedure, or it may be based on a system of three time estimates, as described in Chapter 11 which deals with PERT, the statistical approach to project planning. Regardless of which estimation procedure is used, the scheduling computations described in this chapter are the same, since they deal only with the estimates of the mean activity duration time.

The basic scheduling computations first involve a forward and then a backward pass through the network. Based on a specified occurrence time for the initial network event, the *forward pass* computations give the *earliest (expected) start and finish times* for each activity, and the earliest (expected) occurrence time for each event. The modifier "expected" is sometimes used to remind the reader that these are estimated average occurrence times. The actual times, known only after the various activities are completed, may differ from these expected times because of deviations in the actual and estimated activity performance times.

By the specification of the latest allowable occurrence time for the terminal network event, the *backward pass* computations will give the *latest allowable start and finish times* for each activity and the latest allowable occurrence time for each event. (The modifier, "expected," is usually not added to the latest allowable times; however, it would be appropriate for the same reasons given above for the earliest times.) After the forward and backward pass computations are completed, the slack (or float) can be computed for each activity, and the critical and subcritical paths through the network determined.

As mentioned in Chapter 3, it is often appropriate to adopt one working day as the unit of time. It is also convenient to estimate the activity performance times and to make the network computations in working days, beginning with zero as the starting time of the initial project event. The conversion of these computational results to calendar dates merely requires the modification of a calendar wherein the working days are numbered consecutively from a prescribed calendar date for the start of the project. This procedure is discussed further in Chapter 5. For convenience, this chapter will use *elapsed* working days for discussion purposes; it should be understood, of course, that time units other than working days may be used with no changes in the computation procedures. In addition, it is assumed at the start that the project begins at time zero and has only one initial and terminal event. These assumptions will be relaxed later in this chapter.

COMPUTATION NOMENCLATURE

The following nomenclature will be used in the formulas and discussion which describe the various scheduling computations; for brevity, the modifier "expected" has been omitted from all of these definitions of time and slack. Also, these definitions and subsequent formulas will be given in terms of an arbitrary activity designed as $(i - j)$, i.e., an activity with predecessor event i, and successor event j.

D_{ij} = estimate of the mean duration time for activity $(i - j)$

E_i = earliest occurrence time for event i

L_i = latest allowable occurrence time for event i

ES_{ij} = earliest start time for activity $(i - j)$

EF_{ij} = earliest finish time for activity $(i - j)$

LS_{ij} = latest allowable start time for activity $(i - j)$

LF_{ij} = latest allowable finish time for activity $(i - j)$

S_{ij} = total slack (or float) time for activity $(i - j)$

FS_{ij} = free slack (or float) time for activity $(i - j)$

T_s = scheduled time for the completion of a project or the occurrence of certain key events in a project.

FORWARD PASS COMPUTATIONS

As stated above, the purpose of the *forward pass* is to compute the *earliest start and finish times* for each activity in the project on an elapsed working day basis. To get the ball rolling, an arbitrary earliest start time must be assigned to the (single) initial project event. A value of zero is usually used for this start time since subsequent earliest times can then be interpreted as the project duration up to the point in question. The forward pass computations then proceed by assuming that *each activity starts as soon as possible,* i.e., as soon as all of its predecessor activities are completed. These rules are summarized below.

Forward Pass Rules—Computation of Early Start and Finish Times

RULE 1. The initial project event is assumed to occur at time zero. Letting the initial event be denoted by 1, this can be written as:

$$E_1 = 0$$

RULE 2. All activities are assumed to start as soon as possible, that is, as soon as all of their predecessor activities are completed. For an arbitrary activity $(i - j)$ this can be written as:

ES_{ij} = Maximum of EF's of activities immediately preceding activity $(i - j)$

RULE 3. The early finish time of an activity is merely the sum of its early start time and the estimated activity duration. For an arbitrary activity $(i - j)$ this can be written as:

$$EF_{ij} = ES_{ij} + D_{ij}$$

The above rules are applied to the simple network shown in Figure 4-1a. In the forward pass section, Figure 4-1b, the initial project event 1 is placed at 0 on the time scale according to the first rule. Starting with $E_1 = 0$, the early start time of activity 1-2 is 0, and the early finish time is merely

$$EF_{1,2} = ES_{1,2} + D_{1,2} = 0 + 2 = 2$$

The early start and finish times of activities 2-3, 2-4, and 3-5 are determined in a similar manner.

The crux of the forward pass computations occurs at the merge event 5, where it is necessary to consider the early finish times for predecessor activities 3-5 and 4-5 to determine the early start time for activity 5-6, i.e.,

$$ES_{5,6} = \text{Maximum of } (EF_{3,5} = 9 \text{ and } ES_{4,5} = 7) = 9$$

Finally, the early finish time of the final network activity is

$$EF_{5,6} = ES_{5,6} + D_{5,6} = 9 + 3 = 12$$

Thus, the early finish time for the entire project, corresponding to the earliest occurrence time of the project terminal event 6, is denoted by $E_6 = EF_{5,6} = 12$.

The forward pass network in Figure 4-1b has been drawn to scale on a time base, not only as a convenient means of showing the earliest start and finish times for each activity, but also to show the longest path through the network. It should be added, however, that this graphical procedure for making the basic network computations is introduced here for illustrative purposes and is not recommended for routine application. In Figure 4-1b, activities 1-2-3-5-6 form the longest path of $2 + 4 + 3 + 3 = 12$ days duration. The path consisting of activities 2-4-5 has two days of slack, as will be discussed below; this slack is indicated by the dashed portion of the activity 4-5 arrow.

BACKWARD PASS COMPUTATIONS

The purpose of the *backward pass* is to compute the *latest allowable start and finish times* for each activity. These computations are precisely a "mirror image" of the forward pass computations. First, the term "latest allowable" is used in the sense that the project terminal event must occur on or before some arbitrarily scheduled time, which will be denoted by T_S. Thus, the backward pass computations are started rolling by arbitrarily specifying the latest allowable occurrence time for the project terminal event. If no scheduled date for the completion of the

(a) Basic network with activity duration times

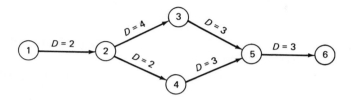

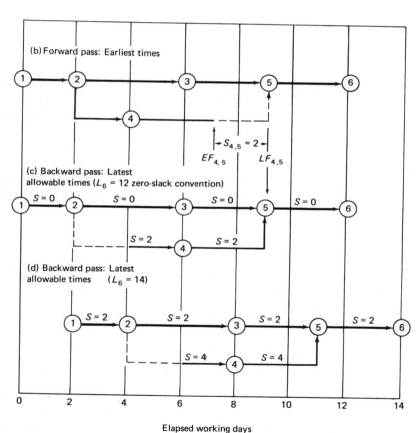

Figure 4-1 Example of forward and backward pass calculations for a simple network.

project is specified, then the convention of setting the latest allowable time for the terminal event equal to its earliest time, determined in the forward pass computation, is usually followed, i.e., $L = E$ for the terminal event of the project. This was followed in the initial development of CPM, and will henceforth be referred to as the *zero-slack* convention.

One result of using this convention is that the slack along the critical path(s) is zero, while the slack along all other paths is positive. This is not true when an arbitrary scheduled date is used for the project terminal event. In this case the slack along the critical path may be positive, zero, or negative, depending on the magnitude of T_S relative to the earliest occurrence times for the terminal event. Following the zero-slack convention, one can also interpret the *latest allowable activity finish time as giving the time to which the completion of an activity can be delayed without directly causing any increase in the total time to complete the project.*

The zero-slack convention is adopted in the illustrative example shown in Figure 4-1c, where event 6 is placed at time 12, i.e., $L_6 = E_6 = 12$. The latest allowable finish time for activities other than the final activity(s) are then determined from network logic, which dictates that an activity must be completed before its successor activities are started. Thus, the latest allowable finish time for an activity is the smallest, or earliest, of the latest allowable start times of its successor activities. Finally, the latest allowable start times for an activity is merely its latest allowable finish time minus its duration time. These rules are summarized below.

Backward Pass Rules—Computation of Latest Available Start and Finish Times

RULE 1. The latest allowable finish time for the project terminal event (t) is set equal to either an arbitrary scheduled completion time for the project, T_S, or else equal to its earliest occurrence time computed in the forward pass computations.

$$L_t = T_S \text{ or } E_t$$

RULE 2. The latest allowable finish time for an arbitrary activity $(i - j)$ is equal to the smallest, or earliest, of the latest allowable start times of its successor activities.

$$LF_{ij} = \text{Minimum of } LS\text{'s of activities directly following activity } (i - j)$$

RULE 3. The latest allowable start time for an arbitrary activity $(i - j)$ is merely its latest allowable finish time minus the estimated activity duration time.

$$LS_{ij} = LF_{ij} - D_{ij}$$

These rules are applied in Figure 4-1c labeled Backward Pass. Starting with the final project event 6, we see that according to the zero-slack convention, it is placed at time 12 which is the earliest time for event 6 computed in the forward pass calculations.

We next compute the latest allowable start time of activity 5-6 by applying rule 3, i.e., $LS_{5,6} = LF_{5,6} - D_{5,6} = 12 - 3 = 9$. The crux of backward pass computations occurs at the burst event 2. Here, the computation of the latest allowable finish time of activity 1-2 requires consideration of its two successor activities. Applying rule 2 above we obtain

$$LF_{1,2} = \text{Minimum of } (LS_{2,3} = 2 \text{ and } LS_{2,4} = 4) = 2$$

Finally, we obtain $LS_{1,2} = LF_{1,2} - D_{1,2} = 2 - 2 = 0$. This result can be used as a check on the computations when the zero-slack convention is followed. If $L_6 = E_6 = 12$ for the terminal event, then $L_1 = E_1 = 0$ must result for the initial event.

DEFINITION AND INTERPRETATION OF SLACK (FLOAT)

Among the many types of slack defined in the literature, two are of most value and are discussed in this text; they are called total activity slack, or simply total slack, and activity free slack, or simply free slack. Total slack and free slack are also referred to by some authors as total float and free float, their definitions being the same as those given below. These definitions of slack are differences between two points in time; they therefore represent intervals of time. Each type of slack has a different interpretation and application as described below.

Total Activity Slack

Definition:

Total activity slack is equal to the difference between the earliest and latest allowable start or finish times for the activity in question. Thus, for activity $(i - j)$, the total slack is given by

$$S_{ij} = LS_{ij} - ES_{ij} \text{ or } LF_{ij} - EF_{ij}$$

The total slack denotes the amount of time (number of working days) that a slack path is away from becoming critical. It is the amount of time by which the actual completion time of an activity can exceed its earliest expected completion time without causing the duration of the over-all project to exceed its *scheduled completion time.* When the convention of letting $L = E$ for the terminal event is followed, total activity slack is equal to the amount of time that the activity completion time can be delayed without affecting the earliest start or occurrence time of any activity or event on the *network critical path,* which is equivalent to not causing any delay in the completion of the project. For example, in Figure 4-1c, the total slack activities 2-4 and 4-5 is two days in each case. Thus, the slack path (2-4-5) is two days away from becoming critical. Consider now what happens if activity 2-4 "slips" by starting late, or by taking longer to complete than the expected duration of two days. For example, suppose activity 2-4 slips one day, and thus its completion time occurs at the end of the fifth day instead of the fourth day, as shown in Figure 4-1b. In this case, the occurrence time for event 4 slips one day, and the total slack for activity 4-5 is thereby reduced by one day. Similarly, if activity 2-4 uses up all of its total slack, i.e., two days, then activity 4-5 will become critical, since its total slack is then reduced to zero. If the completion of activity 2-4 slips more than two days, the duration of the project will be increased accordingly.

Activity Free Slack

Merge point activities (last activity on a slack path) which lie along slack paths have what is called activity free slack. Although this slack concept is not widely used today, it will be defined below and discussed briefly.

Definition:

Activity free slack is equal to the earliest start time of the activity's successor activity(s) minus the earliest finish time of the activity in question. Thus, for activity $(i - j)$, the free slack is given as follows, where $j - k$ denotes a successor activity to the activity in question.

$$FS_{ij} = ES_{jk} - EF_{ij}$$

Activity free slack is equal to the amount of time that the activity completion time can be delayed without affecting the earliest start or occurrence time of *any* other activity or event in the network. The concept of activity free slack will be discussed later in this chapter.

CRITICAL PATH IDENTIFICATION

Now that the concept of slack has been described, the critical path through a network will be formally defined as follows.

Definition:

The critical path is "the path with the least total slack." If the zero-slack convention of letting $L_t = E_t$ for the terminal network event is followed, the critical path will have zero slack; otherwise, the slack on the critical path may be positive or negative. If the network has single initial and terminal events and no scheduled times are imposed on intermediate network events, then the critical path is also the longest path through the network.

For the network in Figure 4-1c, the critical path is 1-2-3-5-6. It is also the longest path through the network in this case. Its duration is equal to 12 days, and it has zero slack since the convention of letting $L_6 = E_6 = 12$ was followed.

To illustrate the case where the zero-slack convention is not followed, it has been assumed in Figure 4-1d that the scheduled completion time of the project is 14 working days, i.e., $L_6 = T_S = 14$. In this case, we see that the critical path remains the same, i.e., 1-2-3-5-6, since it is still the path with least slack. However, the slack along the critical path is now positive, that is, two days, while the slack along the path 2-4-5 is now four days. In this case, a slippage up to two days along the critical path will cause the critical path events to slip a corresponding amount. However, the critical path activities will not slip beyond their latest allowable start and finish times, and in particular, the project end event 6 will not slip past its scheduled completion time of $T_S = 14$ days.

USE OF SPECIAL SYMBOLS IN SCHEDULING COMPUTATIONS

Although there are a number of obvious advantages to having the network drawn to scale on a time base as shown in Figure 4-1, the disadvantages, notably the inflexibility to incorporating network changes, precludes the general use of this procedure. It has been found best in practice not to attach any special significance to the length of the network arrows, but rather to denote the various activity and event times of interest by numerical entries placed directly on the network. As an aid to making the scheduling computations which give these numerical entries, and to display them in an orderly fashion so that they can be easily interpreted, the authors have developed a system incorporating

Reading earliest expected and latest allowable activity start
and finish times and slack from the special symbols

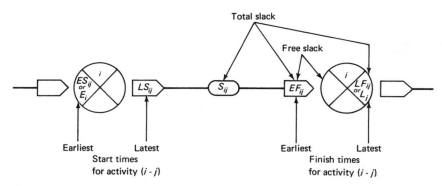

Figure 4-2   Key to use and interpretation of special activity and event symbols.

special symbols.[1] These symbols provide spaces on arrows and in event nodes for recording computed times. The spaces are located logically to bring activity earliest finish (and latest allowable start) times close together and thereby facilitate the computation of earliest expected (and latest allowable) event times.

The use of these symbols is shown in Figure 4-2. The event identification number is placed in the upper quadrant of the node. The earliest event occurrence time, E_i, which is equal to the earliest start time (ES_{ij}) for activity $i - j$, is placed in the left hand quadrant of event i. The latest allowable event occurrence time, L_j, which is equal to the latest allowable finish time (LF_{ij}) of activity $i - j$, is placed in the right hand quadrant of event j. The lower quadrant might be used later to note actual event occurrence times if desired. For each activity, the earliest time the activity is expected to be finished, EF, is placed in the arrow head and the latest allowable start time, LS, in the arrow tail. The estimated duration of the activity along with a description of the activity is placed along the arrow staff. Total activity slack or float is placed in the bubble along the arrow staff. The lower portion of Figure 4-2 points out how one reads the earliest expected and latest allowable start and finish times for an activity. The detailed steps involved in carrying out the scheduling computations using these symbols is illustrated in Figures 4-3 and 4-4.

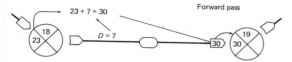

Forward pass

Begin with zero for the earliest start time for the initial project event and compute earliest finish times for all succeeding activities. For a typical activity, place its earliest start time (say, 23 days from project start) in the left quadrant of the event symbol. Then add its duration (7) to the earliest start time to obtain its earliest finish time (30). Write 30 in the arrow head.

Where activities merge, insert in the left quadrant of the event symbol the largest of the earliest finish times written in the arrowheads of the merging activities.

Backward pass

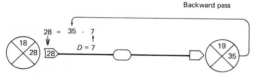

Place the scheduled completion time for the final event in the right quadrant of the project terminal event symbol. For other events, insert instead the latest allowable event occurrence time. For a typical activity, subtract its duration (7) from the latest completion time (35) to obtain the latest allowable activity start time (28). Write 28 in the arrow tail.

Where two or more activities "burst" from an event, insert in the right quadrant of the event symbol the smallest of the latest allowable activity start times.

Figure 4-3 Steps in scheduling computations using special activity and event symbols.

ILLUSTRATIVE NETWORK EMPLOYING SPECIAL SYMBOLS

To further illustrate the use of the special activity and event symbols in making the scheduling computations, a network containing eleven activities is shown in Figure 4-5, which contains the complete forward pass computation. In Figure 4-6 the backward pass and slack computations have been added. With a little practice, the computations made on this network can be completed in two to three minutes. With an allowance for an independent check on these computations, the total time is still less than that required to fill out the ordinary computer input forms. It is interesting to note that this holds true for networks of any size. For this reason, these scheduling computations can be performed more economically by hand than by the computer as long as frequent updating of the network is not required.

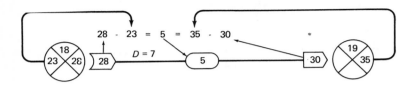

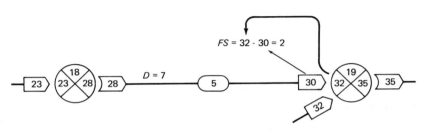

Figure 4-4   Steps in slack computations using special activity and event symbols.

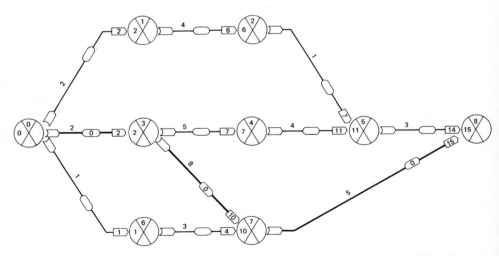

Figure 4-5   Illustrative network employing the special activity and event symbols
showing forward pass computations only.

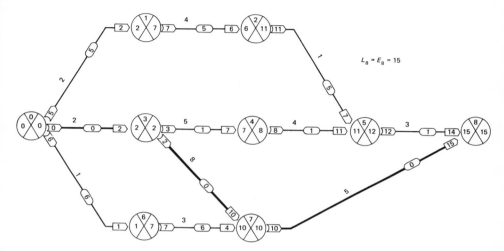

Figure 4-6 Illustrative network employing the special activity and event symbols showing completed computations.

In Figure 4-6, the earliest and latest allowable activity start and finish times are clearly displayed to aid in the making of resource allocation checks, the determination of activity schedules, the conducting of time-cost trade-off studies, etc. For example, consider activity 2-5. Its earliest start and finish times are readily observed to be 6 and 7, respectively, and its latest allowable start and finish times are 11 and 12, respectively. The total activity slack is $S = LF - EF = 12 - 7 = 5$ days, while the activity free slack is only $FS = E - EF = 11 - 7 = 4$ days. This activity illustrates quite well the basic difference between total slack and free slack, the latter occurring only at the end of a slack path, i.e., at a merge event. If the completion of activity 2-5 is delayed up to 4 days, the amount of its free slack, *no other activity or event time in the network will be affected.* In particular, the earliest expected time for event 5 remains at 11. If the completion of this activity is delayed by an amount exceeding its free slack, but not exceeding its total slack, then the earliest expected time for event 5 and the early start time for the following activity 5-8 will be increased. However, no critical path activities or events, such as event 8, will be affected. Finally, if the completion of activity 2-5 is delayed by an amount which exceeds its total slack of 5 days, then the project completion time, i.e., the earliest expected time for event 8, will be increased by a like amount.

According to the previous definition of the critical path, it is made up of activities 0-3-7-8 in the illustrative network shown in Figure 4-6. It has the least amount of slack which is zero in this case, because the convention of letting $L_8 = E_8 = 15$ was followed. This is also the longest path through the network. In addition to determining the critical path through the network, we can identify various subcritical paths which have varying degrees of total slack and hence depart from criticality by varying amounts. These subcritical paths can be found in the following way, which is suggestive of how a computer would handle this problem.

(1) Sort the activities in the network by total activity slack, placing those activities with a common total slack in the same group. Order the activities within a group by early start time.

(2) Order the groups according to the magnitude of their total slack, small values first.

(3) The first group comprises the critical path(s) and subsequent groups comprise subcritical paths of decreasing criticality.

Application of the above procedure to the network in Figure 4-6 gives the results shown in Table 4-1 below.

Table 4-1.   Listing of Critical and Subcritical Paths by Degree of Criticality for the Network in Figure 4-6

Activity	Earliest		Latest		Total Slack	Criticality
	Start Time	Finish Time	Start Time	Finish Time		
0–3	0	2	0	2	0	
3–7	2	10	2	10	0	critical path
7–8	10	15	10	15	0	
3–4	2	7	3	8	1	
4–5	7	11	8	12	1	a "near critical" path
5–8	11	14	12	15	1	
0–1	0	2	5	7	5	third most critical
1–2	2	6	7	11	5	path
2–5	6	7	11	12	5	
0–6	0	1	6	7	6	path having most
6–7	1	4	7	10	6	slack

CRITICAL PATH FROM FORWARD PASS ONLY

The above procedure for locating the critical path(s) is based on a knowledge of total activity slack, which requires the backward pass for computation. While this procedure is necessary to find the slack along subcritical paths, *the* critical path(s) can be determined from the results of the forward pass only. This is quite useful in the early stages of planning and scheduling a project, when it is desired to determine the expected project duration, and to determine the critical path activities with a minimum of computation. The following steps which make up this procedure are based on the assumption that the forward pass computations have been completed, and the resulting EF's and E's have been recorded on the network.

(1) Start with the project final event, which is critical by definition, and proceed backwards through the network.
(2) Whenever a merge event is encountered, the critical path(s) follows the activity(s) for which $EF = E$.

To illustrate this procedure, let us trace the critical path of the network shown in Figure 4-5, on which only the forward pass computations have been made. First we start at event 8 for which $E = 15$. The critical path is then along activity 7—8 since $EF = E = 15$ for this activity, while EF is only 14 along the other path, activity 5-8. In this manner, the critical path can be traced next to event 3, and hence to the initial network event 0.

VARIATIONS OF THE BASIC SCHEDULING COMPUTATIONS

The restrictive assumptions underlying the above basic scheduling computations included the following:

(1) The network contained only one initial event, i.e., one event with no predecessor activities.
(2) The network contained only one terminal event, i.e., one event with no successor activities.
(3) The earliest expected time, E, for the initial event was zero.
(4) There were no scheduled or directed dates for events other than the network terminal event.

These assumptions were made because they simplify the computational procedure and its subsequent interpretation, and at the same time they do not seriously restrict the usefulness of the procedure. However, occasions will be pointed out where it would be of some value to relax these assumptions; hence, the required modifications in the computational procedures will be taken up here.

It may happen, for example, that the network under study is only a portion of a larger project, or perhaps one of several in a multi-project operation. In this case, it may be that the earliest time for the initial network event does not occur at time zero, but rather at some arbitrary number of time units other than zero. In such a case, this specified earliest occurrence time, E, for the initial event is merely used in place of zero, and the forward pass computations are made in the conventional manner.

Similarly, one may wish to specify the latest allowable time for some intermediate network (milestone) event to be a time that is arbitrarily specified as the scheduled time or allowed time for the event in question. In this case one uses a scheduled time, T_S, in determining the L value for this event, as described in the example below. The important points in this discussion are summarized below.

Conventions:

A scheduled time, T_S, for an initial project event is interpreted as its earliest expected time, i.e., $T_S = E$ for initial project events.

A scheduled time, T_S, for an intermediate (or terminal) project event is interpreted as its latest allowable occurrence time.

ILLUSTRATIVE NETWORK WITH MULTIPLE INITIAL AND TERMINAL EVENTS

To illustrate how projects with multiple initial and terminal events and scheduled event times are handled, consider the network shown in Figure 4-7. The "main" project in this network has events numbered 101 through 109. This project, which starts with event 101, produces an "end" objective signified by event 109. However, a by-product of this project is that it furnishes an output for a second project whose events are numbered in the 200 series. For example, the objective of the main project might be to develop a new rocket engine, the completion of which is denoted by event 109. Event 210 might signify the delivery of a key component from a second project that has other end objectives. Activity 108-211 might be the preparation and delivery of a report on the testing of this component which took place in the main project. The scheduled occurrence time of event 211 is thus quite important to the second project, and is independent of the main project.

Two events from the second project are pertinent to the main project network shown in Figure 4-7. First, event 210, which initiates an activity preceding event 106, is scheduled to occur 8 units of time after the start of the first project. Hence, if one sets $E = 0$ for event 101, then $E = 8$ for event 210, since the latter is one of the *initial* events in the network. Event 211 of the second project is scheduled to occur 14 units of time

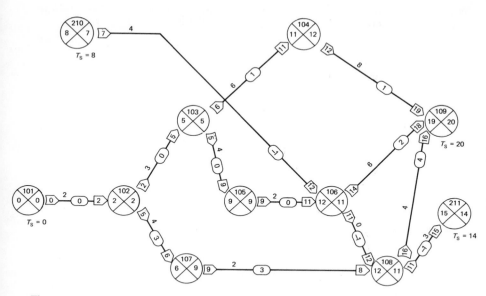

Figure 4-7 Network with multiple initial and end events and scheduled event times.

after the start of the first project. Hence, since event 211 is a *terminal* event in this network, one sets $T_S = L = 14$ for event 211. Finally, the main project is scheduled to be completed in 20 time units, thus $T_S = L = 20$ for event 109. Thus, the network computations are started with the following times specified.

$$\text{Event 101: } E = 0$$
$$\text{Event 109: } L = T_S = 20$$
$$\text{Event 210: } E = T_S = 8$$
$$\text{Event 211: } L = T_S = 14$$

Since early start times are now specified for both of the initial events in this network, the forward pass computations can be carried out in the conventional manner. These computations indicate that the early finish time for the final event of the main project is 18, i.e., $E = 18$ for event 109. The early finish time for the secondary project is 15, i.e., $E = 15$ for event 211. Thus, the schedule can be met on the main project but not on the secondary project, unless the latter is expedited in some way.

Since all terminal events have scheduled completion times, the backward pass computations are initiated by setting $L = 20$ for event 109, and $L = 14$ for event 211. The critical and subcritical paths through the network can then be determined as given in Table 4-2.

Table 4-2.   Critical and Subcritical Paths for the Network in Figure 4-7.

Activities	Total Slack, S	Activities	Total Slack, S
210–106	—1	103–104	1
106–108	—1	104–109	1
108–211	—1	106–109	2
101–102	0	102–107	3
102–103	0	107–108	3
103–105	0	108–109	4
105–106	0		

The critical (least slack) path through the network has a total activity slack of -1. One thus expects to be one unit of time late in meeting the scheduled time of 14 for event 211, the termination of the secondary project. With regard to the main project, one notes that all paths leading to the terminal event 109 have slack of at least one time unit.

Multiproject networks can be quite useful in analyzing the effects of one project on another, and thus offer a means of settling disputes which frequently arise in such a situation. For example, it is clear from Figure 4-7 that one way of alleviating the negative slack situation on the critical path would be to move up the schedule for event 210 so that it occurs on or before the late start time of 7 for activity 210-106. In this way the secondary project could help in solving its own scheduling problem. If this could not be done, then the only remaining remedies would be to reduce the duration of either activity 210-106 or 108-211 by one or more time units. It is also clear from Figure 4-7 that the current schedule for event 210, i.e., $T_S = 8$, does not produce the most constraining path to event 109, and hence does not fix the earliest completion time of the main project.

Another method of handling multiple initial and end events is taken up in exercise 4 at the end of this chapter. This procedure is illustrated in Figure 3-12 where two essentially separate projects are tied together by activity 40-80, which has a time estimate of 40 days. In essence, this activity constrains the completion of one project to precede the other by 40 days. Thus, a scheduled date placed on event 80 in one project can force a scheduled date, or latest allowable time, 40 days earlier on event 40 of the second project. This procedure eliminates the second terminal event.

NETWORK TIME-STATUS UPDATING PROCEDURE

Updating a network to reflect current status is similar to the problem introduced above in that a project underway is equivalent to a project with multiple start events. After a project has begun, varying portions of each path from the initial project event to the end event will have been completed. By establishing the status on each path from progress information, the routine forward pass scheduling computations can then be made as described above. No change in the backward pass computation procedure is necessary, since progress on a project does not affect the network terminal event(s).

To illustrate this updating procedure, consider the network presented in Figure 4-6, which indicates an expected project duration of 15 days. Suppose we have just completed the fifth work day on this project, and the progress is as reported in Table 4-3.

Table 4-3. Status of Project Activities at the End of the Fifth Working Day.

Activity	Started	Finished
0–1	1	3
1–2	4	–
0–3	0	2
3–7	2	–
0–6	2	4
6–7	5	–

NOTE: all times given are at the *end of* the stated working day.

The actual activity start and finish times given in Table 4-3 have been written above the arrow tails and heads, respectively, in Figure 4-8. Events that have already occurred have been cross hatched, and activities that are in progress have been so noted by a flag marked 5 to denote that the time of the update is the end of the fifth working day. Only one of the four paths being worked presents a problem, i.e., activity 3-4. At report time, event 3 has occurred, but activity 3-4 has evidently not started. To avoid this problem, it would be desirable to include in progress reports the intended start time of all activities whose predecessor activities have been completed. If this information is not given, then some assumption must be made to complete the update calculations. The usual assumption, which is the one adopted in Figure 4-8, is that the activity will start on the next working day, i.e., at the end of the sixth working day.

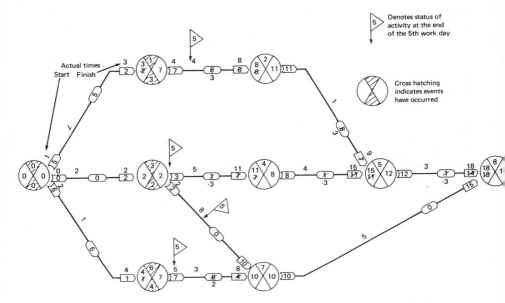

Figure 4-8   Illustrative network showing time status of project.

Having an actual, or assumed, start time for the "lead" activities on each path in the network, the forward pass calculations are then carried out in the usual manner. The original times are crossed out, with the new updated times written nearby. These calculations indicate that the critical path has shifted to activities 3-4-5-8, with a slack of minus three days. Assuming we were scheduled to complete the project in 15 days, the current status indicates we are now three days behind schedule.

NETWORKS WITH SCHEDULED TIMES ON INTERMEDIATE EVENTS

Scheduled times for intermediate network events are handled in a manner similar to the treatment for terminal events as discussed above. In this case, however, there will be two candidates for the latest allowable time for the event in question; the choice is governed by the following convention:

Convention:

The latest allowable time for an intermediate network event on which a scheduled time, T_s, is imposed, is taken as the earlier (smaller) of the scheduled time, T_s, and the latest allowable time, L, computed in the backward pass.

For example, suppose activity 103-105 which involves earth moving, was scheduled to be completed by time 7 to insure completion prior to the ground freezing. In this case L for event 105 would be taken as 7, i.e., the smaller of the scheduled time of 7 and the regular backward pass time of 10. The introduction of this scheduled time reduces the slack on activities 102-103-105 from 1 to -2, a change which is quite important in planning and scheduling these activities. Another interesting consequence of introducing scheduled dates on intermediate events is that the critical path may no longer go through the entire network. In this example, activities 101-102-103-105 have minimum total slack of -2. These activities form a path which starts at the initial project event; however, it terminates at the intermediate event 105 on which the scheduled time was imposed. This is a characteristic result of introducing scheduled times on intermediate events.

SUMMARY OF BASIC SCHEDULING COMPUTATIONS

The basic scheduling computations have been defined as the computation of the earliest and latest start and finish times of each activity; the computation of activity slack then follows immediately. In this chapter a simple procedure has been presented to make these computations directly on the network diagram using special symbols.

Instead of making hand computations directly on the network, one may make them in a tabular manner on a separate sheet. Two tabular procedures are explained in detail in Chapter 7. In comparison with computations on the network, tabular procedures are somewhat tedious and are less efficient than computations made directly on the network. A tabular procedure is included in this text primarily because of its value in helping to understand the logic of computer procedures used to carry out the basic scheduling computations. Anyone planning to program a computer for this purpose should study the tabular procedures in Chapter 7.

The computational problems resulting from the introduction of multiple initial and terminal network events, or the introduction of scheduled times on key milestone events have been treated in this chapter. Their effects on the computational procedures are trivial; however, occasions may arise where they can be profitably applied. Finally, the problem of network time status updating was considered. Again this results in a trivial modification of the basic scheduling computations, but it is an important procedure in the control phase of project management.

REFERENCES

1. Moder, J. J., "How to Do CPM Scheduling Without a Computer," *Engineering News-Record*, March 14 (1963) pp. 30-34.

EXERCISES

1. a. In Figure 4-6, suppose an activity 7-5 must be added to the network which requires 1 time unit to carry out. Will this change any of the times computed in the basic scheduling computations?
   b. What time value for activity 7-5 would cause it to just become critical?
   c. Suppose the project represented by Figure 4-6 is the maintenance of a chemical pipeline in which activity 0–6 represents the deactivation of the line. To minimize the time the line is out of service, when would you schedule this activity?
2. Redraw the network shown in Figure 4-7 and perform the network computations using the following scheduled times.
   a. The main project is scheduled to start at time zero.
   b. The activities leading to event 210 are scheduled to be completed 12 days after the start of the main project.
   c. The scheduled time for the completion of activity 108-211 is 20 days after the start of the main project.
   d. The scheduled time for the completion of the main project (event 109) is also 20 days.
3. a. In exercise 2 what is the critical path?
   b. In exercise 2 what is the effect of assigning a scheduled time of 16 to event 106, or a time of 12 to event 106?
4. It is possible to modify the network in Figure 4-7 so that the correct basic scheduling computations can be carried out by the simple procedure described at the beginning of this chapter. The required modifications are in the form of dummy type activities with suitable time estimates. For example, the addition of an activity 101-210 with a time estimate of 8 would eliminate the multiple initial events. If the main project had a scheduled completion time of 17, i.e., $L_{109} = 17$, what activity and time estimate would eliminate the multiple end events?
5. A reactor and storage tank are interconnected by a 3″ insulated process line that needs periodic replacement. There are valves along the lines and at the terminals and these need replacing as well. No pipe and valves are in stock. Accurate, as built, drawings exist and are available. The line is overhead and requires scaffolding. Pipe sections can be shop fabricated at the plant. Adequate craft labor is available.

You are the maintenance and construction superintendent responsible for this project. The works engineer has requested your plan and schedule for a review with the operating supervision. The plant methods and standards section has furnished the following data. The precedents for each activity have been determined from a familiarity with similar projects.

Symbol	Activity Description	Time (Hrs.)	Precedents
A	Develop required material list	8	—
B	Procure pipe	200	A
C	Erect scaffold	12	—
D	Remove scaffold	4	I, M
E	Deactivate line	8	—
F	Prefabricate sections	40	B
G	Place new pipes	32	F, L
I	Fit up pipe and Valves	8	G, K
J	Procure valves	225	A
K	Place valves	8	J, L
L	Remove old pipe and valves	35	C, E
M	Insulate	24	G, K
N	Pressure test	6	I
O	Clean-up and start-up	4	D, N

a. Sketch the arrow diagram of this project plan. Hint: at least three dummy arrows are required.

b. Make the forward pass calculations on this network, and indicate the critical path and its length.

c. For obvious reasons, activity *E* "Deactivate line" should be initiated as late as possible. What is the latest allowable time for the initiation of this activity?

d. List the various network paths in decreasing order of criticality.

6. The network plan in exercise 5 is subject to criticism because failure to pass the pressure test could result in several problems. How would you network this project to avoid this criticism, and what is its effect on the expected project duration?

7. Update the network given in Figure 4-6 based on the following activity progress report submitted at the end of the fifth working day.

Activity	Start Time	Finish Time	Modifications
0–1	1	3	—
1–2	5	–	—
0–3	0	2	—
3–4	3	–	—
3–7	2	–	—
0–6	5	–	—
5–8	–	–	Activity duration estimate increased to four days for 5-8.

a. What is the current status of this project with respect to a scheduled completion time of 15 days?

b. What activities must be expedited to alleviate the situation found in (a)?

5

USING COMPUTERS
AND CRITICAL
PATH PROGRAMS

In the previous chapter manual methods of making the basic critical path scheduling computations were presented. This chapter will describe how the same computations can be made by computers. The computer programs for CPM and PERT computations are available to anyone who has access to almost any general purpose business (digital) computer. The ready availability of the programs and the ease of use has resulted in the processing of critical path programs by many people who had never before been introduced to electronic data processing.

The treatment in this chapter begins with what the typical critical path program is and how one can make use of it. The actual programs that one may find will vary somewhat in capacity, speed, features provided, and even sometimes in the computational results. Consequently, this text does not attempt to explain how each available program works. Rather, it is a treatment of how they all work in general, and what to look for in the way of good and poor features.

The basic critical path scheduling computations are emphasized in this chapter. The roles of the computer in what might be termed the advanced CPM and PERT topics are mentioned here but are covered more fully in the appropriate chapters of Part II. The problems of actually writing critical path programs, including the full code for one program, are included in Chapter 7.

DEFINITION OF PROGRAM

We shall define a *critical path program* as a computer routine that performs the forward pass, backward pass, and slack computations associated with a CPM, PERT, or similar network technique.

The program is a set of instructions for a particular computer, but it will work for any network (within certain capacity and other limitations, as discussed later). The program may be compared with a square root routine in a desk calculator. The routine is built into the machine, and the user does not really have to know how it works. One must

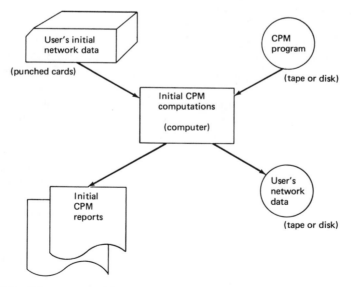

Figure 5-1   Schematic flow diagram of initial CPM processing.

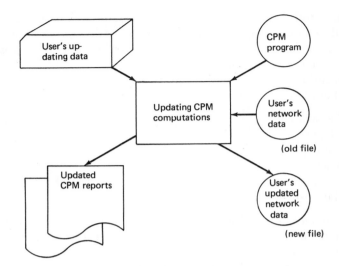

Figure 5-2   Schematic flow diagram of updating computation.

follow certain input procedures to provide the number for which the root is to be computed.

Similarly, a critical path program is associated with a machine. It may actually be in the form of a deck of punched cards or stored on a magnetic tape or disk. To use the program for any network one needs only to follow the proper input procedures in order to set up the particular network for computation. Thus, the *input data* for a given network is the user's responsibility, while the critical path program itself is normally provided by the computer department or service organization.

After the initial computation has been made, the computer may store the user's network on a magnetic tape or disk to await updating computations. A schematic diagram of these operations is given in Figure 5-1. At the time of updating, the user provides input data on the progress of the project and changes in the network. These data are then processed against the initial network the computer has stored. A schematic diagram of this updating function is shown in Figure 5-2.

WHEN A COMPUTER IS NEEDED

In the aerospace and construction industries, contracts that include CPM and PERT requirements for progress reporting purposes often specify that a computer must be employed to do the processing. In most other industrial applications, however, project planners have a choice of manual or computer processing. Although the size of the network is a major factor in deciding whether electronic assistance would be economically or otherwise desirable, size alone should not be the deciding factor. The question involves a number of other factors, among them being:

(1) The availability and cost of a computer with an adequate critical path program.

(2) The expected frequency of updating computations.

(3) Whether cost control, time-cost optimization, resource allocation, statistical or other advanced computations are desired.

(4) The desired formats of the outputs (computational results).

Each potential user must consider these factors in relation to his project and the particular program available to him. A few hypothetical examples of such considerations are given below. These examples are not offered as firm guidelines for the types of situations described, but rather as indications of the considerations that should influence the decision of whether or not to use a computer.

CASE 1 *Heart Surgery.* A network of 75 activities has been prepared for the purpose of planning a complex surgical operation. The network will not be used after the plan has been worked out. *Recommendation:* In this case a manual computation and a time-scaled network would be the best means of working out a well-coordinated plan.

CASE 2 *Promotional Campaign.* A project to introduce a new consumer product involves about 80 advertising, manufacturing, distribution, and sales activities. The network is to be used for initial planning and coordination throughout the six-month project in order to assure proper timing of each phase. All the supervisors have had a short course in CPM. Management intends to update the network twice a month and distribute copies of the results to each of the key supervisors involved. *Recommendation:* In this case the frequency of updating indicates that computer processing would be faster and more economical than manual updating. Also, the computer can sort the output by responsibility, so that each supervisor can get an extracted report on only his activities.

CASE 3 *Construction of a Large Building in a Remote Location.* A five-story building is to be constructed in a small city where no critical path computer program is available. The local contractor is interested in applying CPM in planning the project, though, because this will be his first multi-story building project. The first draft of his network contained 270 activities. *Recommendation:* The manual computation of this network will be well worthwhile, perhaps saving the contractor a great deal of cost and confusion over the re-use of concrete forms, the phasing of carpentry, steel, electrical, and other trades, and by helping to solve other problems of coordinating multi-story construction that he has not yet experienced.

CASE 4 *New Plant Start-Up.* This project includes construction of a plant, delivery and installation of machinery, recruiting and training a new labor force, and starting production. The key supervisors for the contractor, machinery suppliers, training and production departments have had no CPM training, and management considers it impractical to undertake such training at this time. The project manager's network contains 210 activities, and he intends to use it for schedule control during the project. A good CPM program is available. *Recommendation:* The pros and cons of using a computer in this case are not strong either way. The computer reports may be confusing and useless to supervisors not trained in how to read them. On the other hand, experience has shown that machinery suppliers and subcontractors tend to place greater effort on schedule control when the customer shows them periodic computer reports indicating that they will delay the project unless certain tasks are completed by certain dates. Here the psychological factors are more significant than the time or costs of processing. The decision could go either way, depending on the individuals involved.

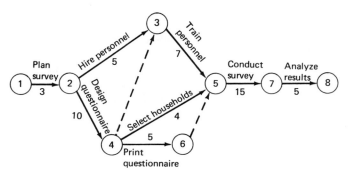

Figure 5-3 Network of survey project.

INPUT FOR INITIAL COMPUTATION

To illustrate how a typical critical path program is utilized, let us take the market survey network in Figure 5-3 and prepare it for computer processing. To do this one must fill out a special input form, dictated by the critical path program being used. Most of the forms are similar to that shown in Figure 5-4. Data related to an activity is entered in one row of this form. The column labeled "pred" is for the predecessor event number of the activity, and "succ" is for the successor event number. The column labeled "t" is for the activity time estimate. (PERT input forms have columns for three time estimates.) The description column usually accepts a certain number of alphabetic or numeric characters, which are used to describe the activity.

The way the form is filled out illustrates one of the rules that must be followed: all dummies must be entered into the form just as all other activities, except that dummies carry zero time estimates. The description "dummy" is optional. Note also that the activities are entered in the form in random order, which is permissible. (We have also assumed that the program to be used will accept events numbered at random; otherwise numbers 3 and 4 in the network would have to be reversed.) Other com-

COMPUTER INPUT FORM

PRED	SUCC	t	DESCRIPTION
2	3	5	HIRE PERSONNEL
1	2	3	PLAN SURVEY
2	4	10	DESIGN QUESTIONNAIRE
4	3	0	DUMMY
6	5	0	DUMMY
4	5	4	SELECT HOUSEHOLDS
5	7	15	CONDUCT SURVEY
7	8	5	ANALYZE RESULTS
4	6	5	PRINT QUSS QUESTIONNAIRES
3	5	7	TRAIN PERSONNEL

Figure 5-4 Basic input for critical path computer programs.

mon rules of input are that each activity must be entered, and no activity may be entered twice. Programs that have scheduled date, calendar date, cost control, and other features have additional columns in the input forms for certain necessary data for these features.

After the form has been filled out and double-checked for errors or omissions, the form is given to a keypunch operator who transfers the information into a deck of punched cards. Each card represents one activity, or one row of the form. (Some programs require so much information that two or three cards are required for each activity.) The punched cards are then fed into the computer system for processing.

Not shown in Figure 5-4 are "header cards" that specify certain general information and parameters for the network. The header cards would normally include space for the name of the network, whether the initial or updating computation is desired, and such other data as required by the options of the program. For example, programs having calendar dating capability normally would require that the header cards include:

(1) the network start date,
(2) the network target completion date,
(3) whether a 5-, 6-, or 7-day week is to be assumed, and
(4) which holidays are to be included.

The requirements for such header data and their input format vary with each program, therefore they are not illustrated here.

OUTPUT FOR INITIAL COMPUTATION

The basic output report contains the same information as given by manual methods. The typical format is shown in Figure 5-5. Note that

MARKET SURVEY PROJECT   **INITIAL SCHEDULE**

I	J	DUR	DESCRIPTION	ES	LS	EF	LF	S
1	2	3	PLAN SURVEY	0	0	3	3	0
2	4	10	DESIGN QUESTIONNAIRE	3	3	13	13	0
4	3	0	DUMMY	13	13	13	13	0
3	5	7	TRAIN PERSONNEL	13	13	20	20	0
5	7	15	CONDUCT SURVEY	20	20	35	35	0
7	8	5	ANALYZE RESULTS	35	35	40	40	0
4	6	5	PRINT QUESTIONNAIRE	13	15	18	20	2
6	5	0	DUMMY	18	20	18	20	2
4	5	4	SELECT HOUSEHOLDS	13	16	17	20	3
2	3	5	HIRE PERSONNEL	3	8	8	13	5

Figure 5-5   Basic computer output report.

the first four fields contain the input data for each activity. The remaining fields provide the computed results for each activity. In this example the times are expressed in units, the unit being defined by the user. In most programs one may elect to obtain the output times in the form of calendar dates.

Note also that the output in Figure 5-5 is now in a particular order, although the activities were put in at random. The order here is by slack, from the lowest to the highest. This "slack sort" is one of the most commonly requested sorts of the output, because it lists the most critical activities first, and thus easily identifies the critical path, or paths.

INPUT FOR UPDATING

To update a critical path schedule after the project has begun and progress has been made, the computer programs may call for the following types of data:

(1) Beginning date for activities that have begun.
(2) Finish date for activities that have been completed.
(3) New data for activities that have been changed.
(4) Complete data for activities that have been added to the network.
(5) Identification of activities that have been deleted from the network.

Most updating programs utilize only start and finish dates of activities, although a few will accept a "percent completion" figure.

For changed activities, normally only the changed field must be entered. This will usually be the time estimate or the description. The *I* and *J* numbers may not be changed, except by deleting the activity and adding a new one with the new numbers. The *I* and *J* numbers are the keys to identification of each activity in the computer files.

Care must be taken in deleting activities, to insure that errors or gaps are not created in the network. If activity 4-3 in the sample network were deleted the computer would accept the input and process it, although the result would be different than when 4-3 is included. However, if activity 3-5 were deleted (and not replaced by one or more other activities) the network would have more than one terminal event; some programs would halt on this condition.

In addition to the above data for the activities, updating input must include some "header cards" as in the input for the initial computation. An important fact for the updating run is the effective or "cut-off" date of the report. Some computations will involve the effective date. Also the effective date serves as the basis for certain error checks. For example, a reported actual date (for the start or finish of an activity) may not be later than the effective date.

To illustrate how typical updating data are entered, consider the survey project of Figure 5-3. Assume that at the end of day 12 progress has been made according to the information in Table 5-1.

Table 5-1.   Progress on Survey Project as of the End of Twelfth Day.

Activity	Actual Start Date	Actual Finish Date	Change
1–2	0	2	
2–4	2	12	
4–6	12	–	
3–5			Change time estimate to 10 days.

Assume that no other progress dates or changes have been recorded as of the end of day 12. These updating data may be entered on the program input form as shown in Figure 5-6. The key to the "Update Code" is as follows: 1= start date; 2 = finish date; 3 = change. Note that only the activities that are actually involved in the updating are entered on the form.

It is emphasized that these procedures are presented only as representative of how some programs handle initial and updating information. The particular program that a user may obtain will probably differ from these examples in a variety of ways. This is especially true of the updating function. Consequently, the user must carefully study the instruction manual provided with the program to be used.

I	J	DUR	DESCRIPTION	Update Code	Date
1	2			1	0
1	2			2	2
2	4			1	2
2	4			2	12
4	6			1	12
3	5	10		3	

Figure 5-6   Input for updating computation.

OUTPUT FOR UPDATED SCHEDULE

To illustrate how updating input may be handled by a computer program, let us assume that day 40 is the established target date for completion of the market survey project, and that the computer program will make its backward pass from this target date in each updating run. (Some programs have this feature, others do not.) Under these conditions the computer output should be basically as shown in Figure 5-7.

The format of the updated report in Figure 5-7 is different from that of the initial report shown in Figure 5-5. The actual duration ("ACT DUR") now appears for the activities that have been completed. This figure is simply the difference between the reported start and finish dates, which are shown under the *LS* and *LF* columns, respectively. The preceding words "STARTED" and "FINISHED" indicate that the dates that follow are the actual reported dates rather than the computed *LS* and *LF* dates. Note also that the finished activities no longer show a slack figure, and are listed first in the slack sort.

The next four activities form the critical path, which has changed since the initial computation. The dummy activities have been automatically deleted from this report because they no longer play a role in the balance of the network. That is, the activities preceding the dummies have been completed, removing the precedence-constraint purpose of the dummies. (Some programs continue to list the dummies anyway. Other programs offer the user the option of omitting all dummies and completed activities from update reports.)

The update report, sorted by slack, provides a straightforward state-

MARKET SURVEY PROJECT   **EFFECTIVE DATE: 12**

I	J	Est Dur	Act Dur	Description	ES	LS	EF	LF	S
1	2	3	2	PLAN SURVEY	STARTED	0	FINISHED	2	
2	4	10	10	DESIGN QUESTIONNAIRE	STARTED	2	FINISHED	12	
2	3	5		HIRE PERSONNEL	13	5	18	10	−8
3	5	10		TRAIN PERSONNEL	18	10	28	20	−8
5	7	15		CONDUCT SURVEY	28	20	43	35	−8
7	8	5		ANALYZE RESULTS	43	35	48	40	−8
4	5	4		SELECT HOUSEHOLDS	13	16	17	20	3
4	6	5		PRINT QUESTIONNAIRE	STARTED	12	17	20	3

Figure 5-7 Update deport.

ment of the project status. One sees at a glance which activities are completed and, of those that remain, which are most critical. The fact that the project is eight days behind schedule is also immediately apparent. The cause of the late condition can be discovered by looking at the records for activities preceding the critical path and the initial activity on the critical path. The reader will see that activity 1-2 started on time and finished a day early. However, activity 2-3, which begins the new critical path, did not start as early as it could have. In fact, as of day 12 it still has not been reported started, meaning that it cannot start until day 13 at the earliest. Furthermore, the estimated duration of activity 3–5 has been changed from 7 to 10 days. For these two reasons, then, the expected completion date of the project is now day 48 instead of the desired day 40.

The slack sort is not the only useful sequence of the output. For large networks an $I - J$ sort is useful for quickly locating data on particular activities. A sort by one of the date columns can also be of value. For example, a sort by the LF column provides a list in chronological order, with the activities that should be finished first listed at the top. All of these sort options and others are usually available through simple "sort key codes" that can be specified by the user. Most users will specify two or three different sorts for each update report.

One of the disadvantages of CPM processing by computer is that the results are separated from the network, necessitating a certain amount of back-and-forth reading between the network and the computer output. Proponents of manual computation directly on the network chart enjoy the ability to read the network and computations simultaneously, which facilitates observation of possible improvements, unexpected resource conflicts, and other aspects of schedule interpretation. (Computer-produced schedules can also be transferred to the network, but only at a much greater expense than doing the computations manually in the first place.)

HOW TO FIND COMPUTER SERVICES

The availability of computer services is a twofold question; first, is a computer available, and second, is an adequate critical path program available for that computer? Organizations not possessing a computer may survey their city or nearby cities for computer installations that rent time on the equipment. The local representatives of the computer manufacturers are good sources for this information. Rental time is offered by many banks and industrial firms as well as computer service companies, and such services are beginning to be found in relatively small cities and towns.

Having determined what equipment is available, one should determine next whether a suitable critical path program is available for that equipment. The organization operating the computer may have a suitable program in its library, the manufacturer may be able to supply one, or it may be necessary to seek a program from other sources. The appendix to this chapter lists a number of CPM and PERT programs, most of which are available to the public. These programs differ significantly in a variety of ways, however, and the differences are worthy of close attention by the potential user.

It is also possible now to obtain network computational service on a time-shared basis. The user provides the input through a rented or purchased terminal, which may be a TWX unit. The input is transmitted over telephone lines to a large computer at some other location, perhaps in another city. The computer processes the input immediately, while it is simultaneously processing a number of other jobs for other users. The critical path output comes directly back to the user's terminal. To obtain this service, the user rents the terminal on a monthly basis and pays only for the computer time actually used. The charge includes the use of the critical path program as well as a variety of other programs available to the user.

Time-sharing is expected to become increasingly popular among small firms that cannot afford their own computers. For the present, however, the most common method of network processing by users who do not have their own computers is to take the data to a computer service bureau for batch processing.

In any case, it is important to obtain and study the "user's manual" that is published for each available program. The manual describes the characteristics of the program and how to provide input for it. When attempting to use a program for the first time, it is a good idea to test the published procedures with a small network of about a dozen activities.

COMPARATIVE FEATURES

In order to summarize the advantages and disadvantages of various critical path programs (the reader is reminded that the term "critical path programs" as used in this chapter refers to programs for CPM, PERT, and similar network methods), certain key features of the programs are listed below and commented upon.

The features discussed are by no means exhaustive of the points on which critical path programs differ. They are among the most significant, however, to the average potential user.

1. *Capacity.* The capacity of critical path programs is usually expressed in terms of the number of activities permitted, but may be in terms of events or both activities and events. Capacities vary from a few hundred to at least 500,000 activities.

   If a network exceeds the capacity of the program, it is not usually practical to divide the network into parts for separate computation. The interaction of forward and backward passes along all paths can make such subdivision highly complex. A better approach is to condense the network, as explained in Chapter. 3.

2. *Event Numbering.* A few of the oldest CPM programs require that events be numbered in ascending order, that is, each activity's successor event number must be larger than its predecessor event number. This is a severe restriction, for it inhibits the flexibility of the network and causes event-number bookkeeping problems. One of the important features of critical path management is the flexibility of the methodology, which facilitates changes in plans and estimates as the project progresses. But if a change means renumbering many events, the change is discouraged and the network tends to become obsolete. The bookkeeping problems also invite errors.

   If events are numbered at random rather than in ascending order, the computer must perform a renumbering function, called "topological sorting" or "ranking," in order to make forward and backward passes in an efficient manner. These ranking routines are challenging to the best computer programmers, and it is little wonder that many of the early critical path programs required that the person drawing the network do his own ranking before the computer would accept the computational task. However, the majority of programs now perform the ranking function and thus permit *random numbering* of events. At least one network procedure, however, does not require topological sorting; see Chapter 7. (Still, a network may not have two different events with the same number, and some manual event-number bookkeeping is necessary to avoid this common error in the use of any critical path program.)

3. *Multiple Initial and Terminal Events.* Critical path programs frequently require that networks have only one initial event and only one terminal event. On the other hand, other programs permit

multiple initial and terminal events. There are interesting pros and cons on this issue.

It may be argued that the single-start-and-end approach imposes a beneficial discipline on the network diagrammer, forcing him to adhere to the logic that every project activity must be initiated by events logically dependent—directly or indirectly—on the starting event of the project, and must be terminated by events leading ultimately to the event marking the project's completion. "Dangling" events or activities (those missing either a predecessor or a successor) are, therefore, regarded as indications of a project plan that is logically untidy.

The counterargument in favor of multiple initial and terminal event capability is based on the fact that accurate and efficient computations sometimes demand such flexibility. Projects may indeed have more than one starting point and/or more than one objective, as discussed in Chapter 4. Where this is the case, the computations are inconvenient and sometimes incorrect if all the dangling events are artificially brought together to single initial and terminal events, which is a procedure frequently recommended. Also, if it is desired to merge two or more parallel projects for a single critical path computation, the multiple initial and terminal event capability greatly simplifies the mechanics of the procedure.

4. *Calendar Dates.* Many programs now have a calendar dating option, which will provide all output dates in the form of 08/25/69 or 25AUG69. To use this option the user needs only to input the base starting date for the first event in the project. A few other options are sometimes available, such as whether the calendar computation is to be based on 5-, 6-, or 7-day weeks, and whether there are holidays. The calendar dating option is highly desirable, although some industrial projects are scheduled by hours and would not use calendar dates. An example of a calendar-dated output is given in Figure 5-8.

The calendar dating routine works by matching the computed time units to a calendar, starting with day 1 on the first calendar date of the project. Weekends and holidays are usually skipped, as illustrated in Table 5-2. The calendar used by the computer may be stored in the program in a tabular form, or each date may be computed by a calendar algorithm (procedure).

5. *Scheduled Dates.* Most programs will accept scheduled dates assigned to the terminal events in the network, and backward

SETUP FOR SAMPLE PROBLEM - COMPUTER DATA CONVERSION PROJECT
MONITOR RUN NO. 2   SAMPLE PROBLEM - COMPUTER DATA CONVERSION PROJECT

SCHEDULED PROJECT FINISH= 17OCT63
EXPECTED PROJECT FINISH= 23OCT63

OPTIONS-

EFFECTIVE 14JAN63   PROJECT STATUS, BEHIND   4 DAYS

ACTIVITY I	J	STATUS	DURATION SCHED	USED	START DATES SCHED	EARLIEST	LATEST	FINISH DATES SCHED	EARLIEST	LATEST	WK/DA FLOAT	EARLY LATE	GAIN LOSS	SL IP
8 000000	9	DUMMY	0/0		19FEB63	21FEB63	18APR63	19FEB63	21FEB63	18APR63	+8/0			
10 000000	12	DUMMY	0/0		08MAR63	28FEB63	26AUG63	08MAR63	28FEB63	26AUG63	+25/0			
11 000000	12	DUMMY	0/0		03APR63	21MAR63	26AUG63	03APR63	21MAR63	26AUG63	+22/0			
11 000000	13	DUMMY	0/0		03APR63	21MAR63	29JUL63	03APR63	21MAR63	29JUL63	+18/0			
0 010003	22	DELIVER COMPUTER	36/0		01OCT62	STARTED	15OCT62	17JUN63	01JUL63	02OCT63	+13/0	-2/0	-2/0	
2 020002	3	MAKE MANUAL VALIDATION PLAN	5/0	4/4				05DEC62	FINISHED	07DEC62		-/2	-/2	
3 020003	4	TRAIN PERSONNEL MAN. RCDS VAL.	1/0	1/0				02JAN63	FINISHED	17DEC62	+2/0	+2/0	-/1	
5 020003	13	VALIDATION FOLLOW-UP	2/0		22FEB63	31JAN63	15JUL63	08MAR63	14FEB63	29JUL63	+23/0			
3 020004	5	SET FOLLOW-UP PLAN	1/0	1/3				02JAN63	FINISHED	14JAN63		-1/3	-3/4	
4 020005	5	MAKE MANUAL VALIDATION	5/0		10JAN63	STARTED	26DEC62	14FEB63	31JAN63	15JUL63	+23/0	+2/0	-1/1	
1 030001	6	HIRE AND TRAIN PROGRAMMERS	6/0 12/0	10/4				27DEC62	FINISHED	20DEC62		+4/4	+4/4	
2 030002	6	MAKE CONVERSION PLANS + SCHED	9/0	9/4				04JAN63	FINISHED	10JAN63		-/4	-/4	

Figure 5-8 Calendar-dated output of CPM monitor. Update program for GE-225 computer. *(Courtesy General Electric Computer Department)*

passes are made from these scheduled dates, rather than merely "turning around" on the terminal event's earliest expected date as computed in the forward pass. The acceptance of scheduled dates means that the slack figures will be related to the scheduled dates, and the critical path may have positive, zero, or negative slack. (The critical path is defined as the path of least slack.) Without this feature, one must perform hand or mental computations to determine the relationship of the expected completion date and the scheduled date.

Table 5-2. Portion of a Working Day Calendar for a Project that Began on January 5, 1969

Working Day	Calendar Date
1	January 5, 1969
2	6
3	7
4	8
5	9
—	Weekend $\begin{cases}10 \\ 11\end{cases}$
—	
6	12
7	13

In some programs scheduled dates are permitted at any intermediate event. In such programs the slack computation is based on either the scheduled date or the computed latest allowed date, whichever is more constraining (earliest). This computation can result in a discontinuous critical path, i.e., one that may begin in the middle of a network.

The use of intermediate scheduled dates frequently results in apparent inconsistencies in managerial logic. To illustrate, visualize a critical path having a total slack of, say, −20 days computed from an intermediate, or milestone event. Assume all other chains of activities, including those leading to the terminal event, have positive slack (are expected to be completed ahead of their directed dates), meaning the over-all project is ahead of schedule. But the steps leading to the intermediate milestone are behind schedule. In this situation an obvious question arises: What is the meaning of the scheduled date for the milestone? In some cases the answer may be clear, such as a scheduled date for the use of a test facility or the anticipated date of spring floods in a bridge project. Too often, however, intermediate milestone dates are set by gross estimation methods prior to the preparation of a detailed network for the project. In such cases the network essentially makes the estimated schedule date obsolete. Furthermore, without a full understanding of the slack figures produced by intermediate scheduled dates, top management may be misled by the critical path reports when they receive them. In practice this kind of misunderstanding frequently occurs. Consequently, the capability to compute slack from intermediate directed dates should be used with due caution.

At least one computer program compromises on this issue by computing the probability of meeting intermediate directed dates and terminal dates, but basing the slack computations only on terminal dates. Other programs provide the option of permitting the scheduled date or the latest allowed date to be controlling on the computation.

6. *Error Detection.*   Critical path programs vary widely in their capability to detect and diagnose network errors and inconsistencies such as loops, nonunique activities, improper time estimates, and excessive terminal events. One of the original PERT programs would not detect a loop at all; it would simply continue to run in a vain effort to perform the topological sort. (With this program, computer operators were instructed to stop the machine if it ran longer than five minutes in the sorting phase.) Most programs will not only detect loops and stop, but will print out the event numbers in the loop. This greatly aids in the manual search for the network or input error, which is usually two different events with the same number. Other types of error are less difficult to detect by programmed routines, and thus most programs will provide adequate detection and diagnosis.

Programs for the updating function also differ considerably in error detection and treatment. An actual finish date for an activity that is earlier than the actual start date will usually be treated as unpermissible, causing a program halt in the input audit phase. But having an activity start or finish before its predecessors have finished is not universally considered an error. Some programs will flag this condition for the output report, but will continue processing by essentially ignoring the unfinished predecessors. Although this treatment contradicts the strict precedence logic of the network, some users feel that it is more realistic. The full completion of a predecessor is not always necessary before its successors can begin, and the programs that are more permissive in this respect could be considered "easier to use." On the other hand, some programs are so conservative that they are irritating and almost impractical. For example, one program halts and requires input correction and restart, if an actual start or finish date occurs on a Sunday.

7. *Output Sorts.* The way in which critical path output data are sequenced, or sorted, affects the utility of the report as a management tool. Some of the sorts currently in use are: (1) by total slack, (2) by successor event number, (3) by earliest expected date, (4) by latest allowed date, (5) by responsibility identification code, and (6) by milestone or key-event code. A few programs provide only one sort; most offer options.

Probably the most generally useful sort is the one by total slack, since this results in a list of paths in order of their criticality. The sort by $I - J$ numbers is handy when searching for input errors. The sort by latest allowed date may serve as a "tickler" file. The sort by responsibility code (such as personnel, department, or subcontractor codes) is useful where interest is limited to certain activities in a network. Where interest is limited to selected events throughout the network, as in the case of high levels of management, a sort by designated key events may save clerical effort in preparing reports.

Many programs also offer options to sort on major and minor keys, such as "*ES* within slack" (or slack major, *ES* minor key).

8. *Report Generator.* Although most programs provide their output in a certain fixed format, at least a few programs permit a high degree of flexibility in format. By means of "report generator" routines, the user may select which information (from a list of 20 to 30 available fields of data) he wishes and may specify the columnar sequence of the fields. Thus, he can tailor the reports to his particular needs, eliminating all undesired data. This feature may be considered a "luxury" item—highly desirable but rarely available.

9. *Updating Facility.* The earliest critical path programs did not provide for network updating by exception. To revise the network in any manner, it was necessary to revise the original input cards and resubmit the entire network deck for computation. There was no updating program as such.

Now almost all of the popular programs provide for updating by exception, as illustrated earlier in this chapter. This feature is not only more convenient and economical, but it facilitates accuracy in determining the updated status of a complex project.

Nevertheless, there are significant differences in how the updating function is handled among the available programs. For example, some programs permit only the input of finish dates of activities; the start dates are ignored. This means that if an activity has started late and is not yet finished, the late condition will not be recognized by the program, and the output will be incorrect to that extent. Also, if only finish dates are accepted, it is not possible to compute and record the actual duration of each activity for comparison with the estimated duration.

A similar but more serious error is to assume that, on a path having had some progress dates reported, the next unreported event date on the path will occur at its earliest time (ES or EF), regardless of the effective date of the input. To understand the effect of this assumption, review the market survey network and the sample progress report in Figures 5-3 and 5-6. Activity 2-3 is pertinent, for it has not started as of the reporting date, day 12. The correct handling of this condition is to compare the ES for the activity with the reporting date, and take the larger of the two as the basis for beginning the forward pass. If the report date is larger, one day may be added. Thus, since 2-3 had not started by the end of day 12, the earliest that it could now start is day 13. Therefore, the correct ES for 2-3 is 13, the EF becomes 18, and the project completion date is extended.

By the incorrect procedure, the ES for 2–3 would be assumed to be 2 (which is impossible since day 2 is now ten days ago), the EF would become 7, and this path of the project would apparently be a day ahead of schedule. Incorrect results of this type have caused misunderstanding and confusion for many users. Programs using this incorrect procedure should be avoided.

10. *Graphical Output.* In discussing reports for management, it is important to understand a law that is as valid as any of Mr. Parkinson's: The higher the level of management, the greater is the demand to see charts rather than tabulated data. Since digital computers do not easily produce charts, this law has caused no end of trouble for designers of data-processing systems for management. Some critical path programs include routines that prepare bar charts of selected key activities, "drawing" the bars on a time scale by repetition of some character (for example, see Figure 5-9). Other programs will plot graphs of the distribution of resources. Very few programs will generate network diagrams directly from the computer. Consequently, critical path analysts wisely tend to concentrate on making the tabular reports as read-

```
LOOP TEST CASE
NETWORK START DATE  2 JUL 62              NETWORK REPORT DATE  2 JUL 62                                    PAGE   1
GRAPH BY DAYS BEGINNING AT  1 JUN 62
PREDECESSOR  SUCCESSOR   ACTIVITY DESCRIPTION
                                                    1111111111222222222233333333334444444444555555555566666666
                                          123456789012345678901234567890123456789012345678901234567890123456

ACTUAL   700     67  JOB NUMBER 614/E
         1100            LATEST FINISH  10 JUL  62                   XXXXSS
         2 JUL 62                                                         A

ACTUAL   1700   546  JOB NUMBER 216/B
         500             LATEST FINISH  11 JUL  62                   XXXSSS
         2 JUL 62                                                         A

ACTUAL   555    129  JOB NUMBER 777/9
         900             LATEST FINISH   7 AUG  62                   XXXSSSSSSSSSSSSSSSSSSSSSSSSSS
         2 JUL 62                                                         A

EXPECTED START  100    214  JOB NUMBER 128/B
                6 JUL 62        LATEST FINISH  16 JUL  62             XXXSSS
                                                                         E

EXPECTED START  500    694  JOB NUMBER 424/G
                6 JUL 62        LATEST FINISH  20 JUL  62             XXSSSSSSS
                                                                         E

EXPECTED START  500    213  JOB NUMBER 127/A
                6 JUL 62        LATEST FINISH  27 JUL  62             XXXXXXXXXSSSSSS
                                                                         E

EXPECTED START  1100   105  JOB NUMBER 212/F
                9 JUL 62        LATEST FINISH  16 JUL  62             XXXSS
                                                                         E

EXPECTED START  1100   342  JOB NUMBER 213/A
                9 JUL 62        LATEST FINISH  27 JUL  62             XXXXXSSSSSSSSS
                                                                         E

EXPECTED START  100    210  JOB NUMBER 165/C
                12 JUL 62       LATEST FINISH  20 JUL  62             XXXSSS
                                                                         E

EXPECTED START  100    714  JOB NUMBER 129/B
                12 JUL 62       LATEST FINISH  27 JUL  62             XXSSSSSSSS
                                                                         E

EXPECTED START  100    621  JOB NUMBER 135/H
                12 JUL 62       LATEST FINISH  31 JUL  62             XXSSSSSSSSSS
                                                                         E

EXPECTED START  400    206  JOB NUMBER 128/C
                18 JUL 62       LATEST FINISH  26 JUL  62             XXXXSS
                                                                         E

EXPECTED START  400    155  JOB NUMBER 129/D
                18 JUL 62       LATEST FINISH  16 AUG  62             XXXSSSSSSSSSSSSSSSSSS
                                                                         E
```

able and useful as possible, using many of the sorting features listed above, and thus avoid to a large extent the expense of manually preparing and maintaining time-scaled graphical displays for management.

11. *Network Condensation.*   Routines have been developed recently which in effect condense large networks into smaller ones. One such procedure involves three phases: (a) the condensation of large, detailed networks into smaller ones, (b) the integration of two or more condensed networks, and (c) the expansion of the condensed and integrated networks back into large, separate, detailed networks. The condensation phase involves the determination of the longest path between each sequential pair of preselected key events in the detailed network. The result of this phase is a network consisting of the key events only, which are connected by single activities instead of groups of activities. (The network here, of course, is a tabular representation in the computer's memory, not a diagram.) In the integration phase the computer processes two or more condensed networks as though they were a single network, each condensed network having one or more common events (interfaces) with another network. The results of this process are new earliest and latest times for each key event. The expansion phase, then, utilizes the new restraints on the interface events to compute new schedules for all activities in the detailed networks. The output of the expansion phase is in the same format as the output for the original detailed network, but the earliest and latest times and slack figures reflect the new restraints imposed on the interfaces.[1] Hence, this network condensation-integration-expansion procedure provides extraordinary capacity for processing the largest networks and groups of networks. It also aids in the preparation of summarized reports for management.

12. *Activity Vs. Event Orientation.*   One may hear of critical path programs as being either "activity oriented" or "event oriented." This classification is intended to imply that the input and output data, especially descriptions and scheduled dates, are associated either with activities or with events. The distinction is not a substantive one with respect to computational procedures. The input and output formats differ somewhat with the orientation, and evaluation of these formats is largely a matter of personal preference. Activity orientation appears to be the more natural form of expression, although milestone reporting is somewhat easier with event orientation. Often the difference between the

two is simply the tense of the verb in the descriptor; for example, if an activity descriptor is "perform test," the corresponding event descriptor might be "test performed." Activity orientation predominates in the programs now available. Some programs provide both activity and event orientation.

The node and precedence diagramming schemes (described in Chapter 6) do not involve events. Programs written especially to handle these schemes are quite different in that only one number is used to identify an activity. Node and precedence programs usually are designed to accept either arrow or node notation, but programs based on the arrow scheme will not accept node or precedence input.

13. *Statistical Analysis.*   Originally the controversy of the three-estimate probabilistic approach versus the single-estimate deterministic approach to critical path computations provided a clear distinction between PERT and CPM. While the distinction still exists among many critical path specialists and is maintained in this text, the two terms have been used loosely in practice and are no longer mutually exclusive. For example, some programs permit three time estimates in the input and obtain the usual weighted average of these figures, but do not make any summations of variance or computations of probability. Another system used by a government agency is labeled "PERT" but makes no use of three time estimates or statistics. Consequently, if one desires the probabilistic approach he must be careful not to rely on the title "PERT" or the three-estimate input, but must determine exactly what probabilities are computed by the programs in question. The most common probability computation is an approximation to the probability of meeting a scheduled date, usually the scheduled date for the completion of the project.

Those who are interested in the CPM (deterministic) approach but who find only PERT (probabilistic) programs available to them do not have a serious obstacle. Almost any probabilistic program may be used as a deterministic one by simply entering the single (expected) time estimate as each of the three estimates.

14. *Time-Cost Trade-Off.*   The early CPM programs included a routine which utilized time and cost data for each activity in order to seek an optimal balance between the total project time and cost. Routines based on this concept are available in several programs. The theory of time-cost trade-off is presented in Chapter 9.

15. *Cost Control.*   Cost control features may be defined as those dealing with summarizations of budgeted and expended funds by

time period, based on input data related to activities or groups of activities in a network (see Chapter 10). Models of this type are sometimes called "enumerative," as opposed to the optimization models described above. Some of the existing programs with cost control features are designed for compatibility with the accounting procedures of a particular organization, while others attempt to be more general. Often these models are called PERT Cost models. Cost control procedures should also provide for various levels of cost summarization and breakdowns by subcontractor, project phase, etc., in order to fully exploit the extra input data required.

16. *Resource Allocation.* In addition to costs, project managers must be concerned with the proper utilization of men, equipment, and facilities to avoid overloads and idle periods that cause delays. A few computer programs have been written which schedule the available resources to avoid overloads and minimize delays. Some of these programs will operate on only one resource category (skill or equipment type) at a time, although some will operate on a large number of resource categories. Other programs will summarize labor or equipment loading by time period, but will not adjust the schedule to obtain an improved pattern of resource requirements by time period. Most resource allocation programs are proprietary. For further discussion, see Chapter 8.

SUMMARY

The question of when and how to use computer programs for CPM computation has been covered in this chapter, along with examples of input procedures and output reports. It is stressed that CPM programs are readily available to the public at most computer centers, and that they are not difficult to learn to use, even for persons totally unfamiliar with data processing. There are some significant differences in the features available in the programs, and the user should study these carefully before making a choice. Some features concern the more advanced aspects of critical path methods, and these are dealt with in Part II.

REFERENCES

1. Prostick, J. M., "Network Integration, a Tool for Better Management Planning," Presented at Meeting of Operations Research Society of America, Philadelphia, November 7–9, 1962.
2. Phillips, Cecil R., "Fifteen Key Features of Computer Programs for CPM and PERT," *Journal of Industrial Engineering*, January–February 1964.

APPENDIX 5-1
LIST OF
COMPUTER PROGRAMS

It is estimated that well over 100 programs have been written by various organizations and individuals for CPM and PERT processing.[2] Many of them have been held as proprietary by the originating firms, but most have been available to the users of the appropriate computer equipment. The earliest programs were written for computer models that are no longer generally in use (such as the IBM 650), and these are being replaced by programs written for the newer machines. As the new programs are developed, some old "standard" features are being dropped and are being added.

To give the reader an idea of what programs and features are available now, a selection of the current offerings is listed in Table 5-3. The list is a result of a

Table 5-3.   Selected List of Critical Path Programs

Computer	Program Name	Capacity	Network Scheme	Updating	Sched. Dates	Calendar Dates	Optional Sorts	Bar Charts	Cost Control	Time-Cost	PERT	Res. Allocation	Comment
Burroughs B200/ B300	PERT/Time	900 events	arrow	X	X	X	X						
Burroughs B5500 and Others	Time-PERT in ALGOL 60	524, 288 activities	arrow	X	X	X	X				X		
Burroughs B5500 and Others	PROMIS Time Module	(not given)	arrow	X	X	X	X				X		Has report generator and other unusual features.
Control Data 1604	PERT	3000 activities	arrow	X	X	X	X				X		
GE-115	Critical Path Method Program	350-3000 events, depending on configuration	arrow					X					
GE-215/225/235	Project Monitor and Control Method (CPM/ PROMOCOM)	999 events	arrow	X	X	X	X			X			Was one of the original updating programs.
GE-400/600	Critical Path Method Program and CPM/ MONITOR	Approx. 3000 events	arrow	X	X	X	X		X				

Computer	Program	Network size	Network type	Features	Remarks
Honeywell 400/1400 or 800/1800	PERT	3000 activities and up, depending on configuration	arrow	X X X X	
IBM 1401 and System/360	Management Control System	4600 nodes	arrow, node, or precedence	X X X X X X	Proprietary program of McDonnell Automation Company, Houston, Texas.
IBM System/360 (min. 32K, two disks)	Project Control System/360	5000 activities	arrow	X X X X X X	Will summarize resources.
IBM 1130	1130 Project Control System	2000 activities	arrow, node, or precedence	X X X X X X	Will summarize resources.
IBM System/360 (min. 64K, two disks, three tapes)	Project Management System (PMS/360)	Depends on configuration and network features	arrow	X X X X X	Modular set of programs. Has multiple network capability and report generator.
NCR-304 or NCR-315	PERT	5000 activities	arrow	X X X X
RCA 501	PERT	2000 activities	arrow	X X X X
UNIVAC 1107	PERT/COST	(not given)	arrow	X X X X X	Will summarize resources.

survey of the major computer manufacturers and other sources conducted by the authors in 1968. The selection of programs in the list is intended to be only representative of the better programs known to the authors. All of the programs listed make the basic critical path computations. Except where noted as proprietary, the programs are available through the computer manufacturers to users of their machines.

The information on each program listed was derived by the authors primarily from the descriptive literature provided on each program. The authors did not actually test each program. The meanings of the features listed in the table are explained below.

1. *Network Scheme.* The program is considered to be based upon the arrow scheme if it uses the customary $I - J$ numbering method. This means that the event scheme may also be used, as explained in Chapter 6. See also Chapter 6 for an explanation of the node and precedence schemes.
2. *Updating.* Programs with updating features will accept actual start and/or finish dates and network changes on an exception basis and will recompute the network schedule. Some programs may make this computation incorrectly, however, as mentioned earlier in this chapter.
3. *Scheduled Dates.* Programs with this feature are not limited to the zero-slack convention described in Chapter 4. Backward pass computations, at least, can be made from scheduled or target completion dates for the network, resulting in positive, negative, or zero slack values.
4. *Calendar Dates.* The output schedules may appear in the form of calendar dates.
5. *Optional Sorts.* The user may specify the sequence in which the activities are to be listed in reports.
6. *Bar Charts.* Some form of bar chart, as illustrated in Figure 5-9, is available as an optional output.
7. *Cost Control.* The program has some provisions for entering a budgeted cost figure for each activity and then the actual cost incurred, and will summarize the budget and actual figures on each updating run.
8. *Time-Cost.* This feature is treated in Chapter 9. It usually requires "normal" and "crash" time and cost estimates for each activity and computes the least-cost means of accelerating the project.
9. *PERT.* This feature is checked only if the program computes a probability of meeting scheduled dates, which is considered the

fundamental distinction of PERT programs. Some programs accept three time estimates and compute their weighted average but do not follow through with the probability index computation. See Chapter 11.

10. *Resource Allocation.* This feature is checked only if the program contains an algorithm that attempts to minimize project duration under certain resource constraints, as discussed in Chapter 8. Programs that will summarize resource requirements by time period are noted in the table under Comments.

TRENDS IN NETWORK PROGRAMS

In the first ten years of critical path method technology the computer programs for network processing have undergone a steady evolutionary development. The early programs that were limited to "*I*-always-less-than-*J*" networks, the zero-slack convention, and provided only one output report, have almost disppeared. While few of the early programs had updating capability, almost all of the latest ones do. Likewise, all the newest programs have calendar dating and scheduled date capability.

It is also interesting to note that the time-cost trade-off feature and the PERT probability computations are not appearing in the latest programs. These two features provided the most significant distinctions in the first CPM and PERT programs. Their lack of sustaining popularity in spite of the increased general use of critical path methods indicates that these features have less practical value than others. Full descriptions of these features are given in Chapters 9 and 11.

Certain other "advanced" features have increased in popularity. Bar charting options are becoming almost standard accessories. Resource allocation procedures, although still presenting extremely difficult problems, are continuing to be developed and made available on a proprietary basis.

Probably the most significant trend has been toward expanded flexibility in the basic scheduling computations and in report formats. The new Burroughs PROMIS program and the IBM PMS/360 system appear to contain high degrees of flexibility, including report generators that permit the user to design the content and layout of his own reports. Such programs are more expensive to develop, and they indicate a growing demand for basic network processing in a wide variety of applications.

Further comments on network programming techniques, primarily from the programmer's point of view, are provided in Chapter 7.

II

ADVANCED TOPICS

6

OTHER NETWORKING SCHEMES AND GENERALIZED NETWORKS

The graphical method of drawing networks that was presented in Chapter 2 and used throughout Part I is called the *arrow* scheme in this text. This scheme was introduced in the original publications on the Critical Path Method. Since then the arrow scheme has remained one of the most common means of portraying networks. The vast majority of computer programs for CPM and PERT are designed to accept the predecessor-successor $(I - J)$ event code for activities that are associated with the arrow scheme.

However, the arrow scheme is by no means the only networking procedure, nor is it the most efficient one if we judge efficiency by the number of symbols required to portray a given number of activities in a network. At least three other schemes have been conceived and used widely. In this chapter these other schemes are described and evaluated. The individual or organization that has a choice of graphic schemes (that is, not limited by contract or customer preference) should consider all of them and select the one most suitable to his purposes. In most cases, of course, it will be desirable that the organization standardize its networks by using only one of the schemes consistently.

In addition, this chapter includes an introduction to the concept that a project network may contain alternative paths which lead to several different end results for the project. This concept raises the question of which paths the project will take, and several scholars have developed techniques of statistical analysis that treat this question. Of particular interest is the perspective provided by these developments, which shows that the usual CPM and PERT networks are only special cases of a more general network concept.

EVENT SCHEME

While the original developers of CPM were drawing arrow networks and writing activity descriptions *along the arrow shafts*, the developers of PERT were also drawing arrow networks; but the PERT developers wrote their descriptions *inside the nodes*, or event symbols. This convention was undoubtedly motivated by the desire to develop a system that was compatible with the milestone system, which has been in use for some time in planning and controlling military programs. Milestones are, of course, key points in time in the life of a project and are, therefore, events.

The developers of PERT thought of the network as a series of activ-

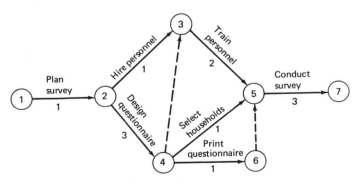

Figure 6-1 Survey network, arrow scheme.

ities that must precede the occurrence of events. Time estimates were applied to activities, but descriptors were in terms of the successor event. Virtually all other networking rules that applied to the arrow scheme also applied to the "event oriented" or the "event" scheme.

To illustrate this scheme, we shall use part of the market survey network used in Chapter 5. Figure 6-1 shows the project in the arrow network format. Figure 6-2 gives the same project in the event format.

In this case the event scheme is noticeably less efficient than the arrow scheme. Although the same real activities are represented, the event network necessarily contains more nodes and more dummies. At merge points, in particular, the event scheme requires more dummies for clear identification of each activity. Otherwise, several activity arrows would converge at a single event descriptor. Then the reader would be unable to discern the arrow that represents the activity described in the event symbol, and which activities are simply not identified. All too often this kind of error occurs in the practical use of the event scheme. As an illustration, see Figure 6-3, which has an alternative representation of the survey project. To a reader not familiar with the project, either activity 3-5 or activity 4-5 could be the "train personnel" activity; and the other activity is not identified at all. Such flaws have been observed frequently in PERT networks, which traditionally use the event scheme.

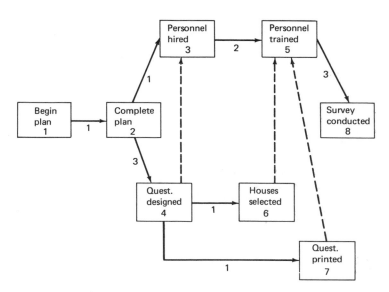

Figure 6-2 Survey network, event scheme.

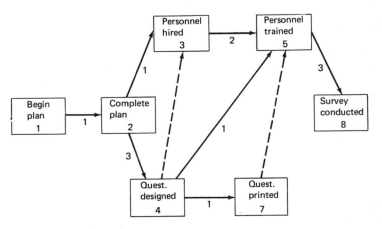

Figure 6-3   Survey network, event scheme with common type of error.

The event scheme also tends to cause misunderstandings about the dependency relationships. In Figure 6-3, for example, "Personnel Hired" appears to be dependent upon "Questionnaire Designed," because of the location of dummy 4-3. Actually, the dummy is a restraint on the start of activity 3-5. While PERT specialists become accustomed to reading the event scheme properly, management personnel and others often misinterpret the logic or ignore it completely.

The only advantage cited by proponents of the event notation is that it is more natural for some managers to read a chart that is in terms of events, which are similar to the "milestones" used in other schedule formats.

The event scheme also facilitates the use of summarizing programs that are based upon the designation of key events (described in Chapter 3), although the same programs will work with the arrow scheme. In fact, as far as computer processing is concerned, the arrow and event schemes are essentially identical. In both schemes the arrow represents the activity and is identified by its I and J event numbers. The $I - J$ identification fully specifies the linkages among the activities.

In summary, the event scheme is a hybrid format for networks, being essentially an arrow scheme with the appearance of a node scheme. It is less efficient and less articulate than either the arrow or node schemes. Although the event scheme is still used widely in the research and development industry, due to the original application of PERT in this field, it is not recommended for users who have a choice of graphic schemes.

BASIC NODE SCHEME

The complete reverse of the arrow scheme is the node scheme, in which the nodes represent the activities and the arrows are merely connectors. The market survey project in the node format is illustrated in Figure 6-4. The principal advantage of the node scheme is that it eliminates the need for special dummies to correct false dependencies. This feature makes the scheme more efficient and, more importantly, easier to learn. In the arrow and event schemes the most difficult aspect is learning to make the proper use of dummies. In the node scheme all the arrows are dummies, in effect, and there are no subtle false dependency problems requiring the use of special dummies. The node relationships also avoid the frequent confusions of the event scheme.

The only practical disadvantage of the node scheme has been the fact that very few computer programs have been written to accommodate the scheme. Programs designed for arrow and event schemes will not accept the node notation, which contains only one identifying number for each activity. The linkages between activities must be input to node-scheme programs by listing all the activities that precede (or succeed) each activity. In recent years, however, several node-scheme programs for widely used computers have been made available (see Appendix to Chapter 5), and the node scheme is apparently increasing

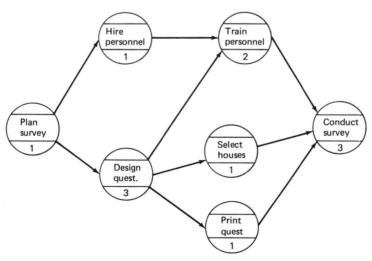

Figure 6-4 Survey network, node scheme (shape of node symbol is not significant).

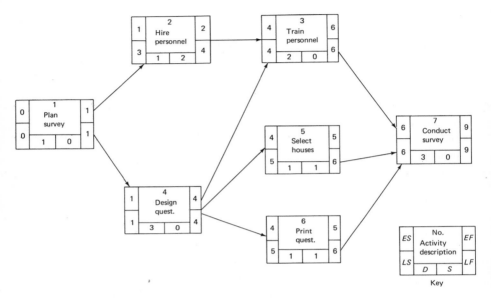

Figure 6-5   A manual computational method for the node scheme.

in popularity among network users. The problems of programming the basic network computations for the node networking scheme are taken up in the next Chapter.

The node scheme lends itself to manual computation as well as the arrow scheme. One symbolic method for manual computation of node networks is illustrated in Figure 6-5. Using the key to the symbol notation, the reader can easily see how the computation is made. For example, the *EF* for node 3 is obtained by adding its times for *ES* and *D*. Thus, $EF = 4 + 2 = 6$. At merge points, such as node 7, the *ES* is selected as the largest of the preceding *EF* times. The backward pass is made similarly.

In comparison with the symbols used in manual computation of arrow networks (as presented in Chapter 4), the node symbols are somewhat more articulate. All of the numbers associated with an activity are incorporated in the one node symbol for the activity, whereas the arrow symbols contain each activity's data in the predecessor and successor nodes, as well as on the arrow itself.

One of the first proponents of the node scheme was J. W. Fondahl of Stanford University.[1] It appears to be a generally superior scheme, especially for manually processed networks. It is perhaps unfortunate that the originators of CPM and PERT hit upon the arrow and event schemes first, and most of the computer programs were designed for the $I - J$ notation. The second decade of critical path technology may see a general conversion of most programs and manual methods to the basic node concept and its variations.

PRECEDENCE DIAGRAMMING

An extension of the original node concept appeared around 1964 in the user's manual for an IBM 1440 computer program. One of the principal authors of the technique was J. David Craig of the IBM Corporation, who referred to the extended node scheme as "precedence diagramming."[2]

The modifications introduced by precedence diagramming consist of additional ways of displaying the dependency relationships between activities. Instead of strictly discrete dependencies (as expressed in Rule 1 of Chapter 2), the precedence rules permit more liberal specification of lags and delays among activity start and finish constraints.

As an example of how the precedence rules apply, consider the project of digging and pouring footings used in Chapter 3. The digging and pouring activities were planned to proceed basically in parallel, although the digging would have to begin about a day before the pouring could begin. At the end, pouring would continue for about a day after digging had been completed. Carried out in this manner the project was estimated to take about 4 days.

The original network representation in Figure 6-6 was clearly incorrect since it shows a six-day duration. Consequently, the activities were

Figure 6-6 First draft of network for digging and pouring footings.

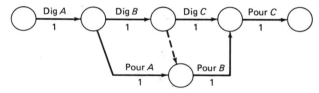

Figure 6-7 Correct arrow network for footings project.

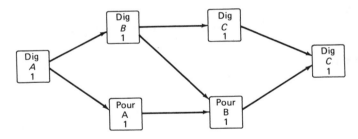

Figure 6-8   Node network.

divided into parts of approximately one day's work each. Then the network in Figure 6-7 could be constructed, which provides a more accurate representation of the four-day project plan. The same project in the node format is shown in Figure 6-8.

A precedence diagram for the project could look like Figure 6-9. The one-day lag in starting the pouring work would be part of the computer input for the pouring activity, and the computational result would be the same total project duration of 4 days. Note that only two nodes and two arrows are required, indicating that precedence diagramming can be a considerably more efficient scheme where such series-and-parallel relationships exist. Another advantage is that each of the two major activities, digging and pouring, are shown as single activities.

To the superintendent in the field, this network corresponds more closely to his own concept of the job, which is that of essentially two activities, rather than six or seven activities and dummies. Therefore, advocates of the precedence scheme feel that it is easier to understand, and it lends itself more readily to "percent completion" progress reports. The field superintendent can simply report 33 percent completion when a third of the digging is done, which is a natural form of estimating and reporting, rather than having to say that "the first third of digging is 100 percent complete." This distinction is especially useful where cost reporting is involved. With a precedence network the cost charge number often can be assigned to the one activity it applies to, rather than to several subactivities.

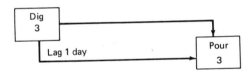

Figure 6-9   Precedence diagram.

Figure 6-10a Finish-to-start relationship. (Start of B depends on finish of A).

Figure 6-10b Start-to-start relationship. (Start of B depends on start of A).

Figure 6-10c Finish-to-finish relationship. (Finish of B depends on finish of A).

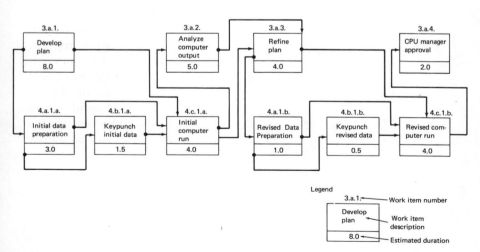

Figure 6-11 Sample network using special precedence notation suggested in IBM publications.[3] *(Courtesy IBM Corporation)*

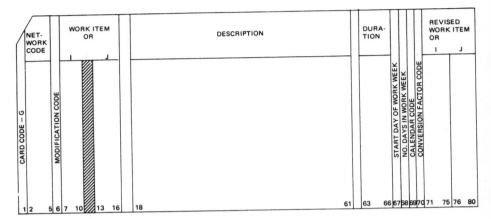

Figure 6-12a   Input card format for basic activity data.

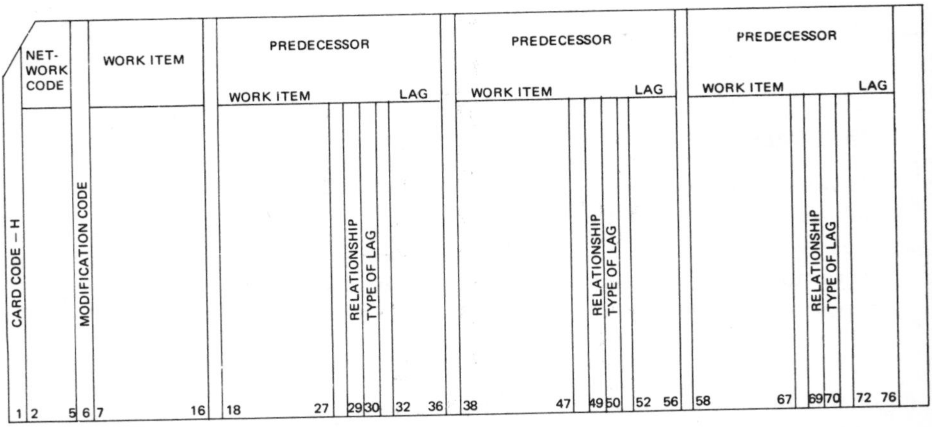

Figure 6-12b   Input card format for activity predecessors. Both cards above are parts of the IBM Project Control System 1360. *(Courtesy IBM Corporation[1])*

A new symbolic discipline is suggested for precedence diagrams in IBM publications.[3] Three different dependency relationships are identified, as shown in Figure 6-10. The end-to-start relationship, of course, is the only relationship used in the arrow, event, and basic node schemes. In the precedence scheme, dual relationships may exist between two activities, that is, where both the start and finish of activity *B* depend upon the start and finish of activity *A* respectively. A sample network using this symbolic scheme is shown in Figure 6-11.

The time lags between activities may be specified in terms of time units or in terms of percentages of the estimated duration of the "base" activity. Identification of the base activity varies with the type of relationship as shown in Figure 6-10.

The types of input required for precedence networks are shown in Figures 6-12a and 6-12b. These are two of the input card types used in the IBM Project Control System/360. The output data is essentially the same as that obtained from programs based on arrow, event, or node networks.

In conclusion, the precedence scheme has much to recommend it in some network situations, especially where activities are interdependent in terms of both starting and finishing, and where close series-and-parallel relationships exist. The precedence scheme adds a significant degree of flexibility and efficiency in these situations, as compared with the basic node scheme. On the other hand, it is obvious that the network planner could get carried away with the variety of symbology available in the precedence scheme, and could easily create such esoteric diagrams that he alone could decipher them. Of course, this would negate the communicative purpose of the network. Therefore, the use of relationships other than the basic end-to-start should be judiciously limited to the portions of the network where they can be used to advantage.

GENERALIZED NETWORKS

In all of the network schemes presented previously in this text, the network logic has been considered deterministic. That is, it was assumed that every path in the network was a necessary part of the project; there were no optional or alternative paths. However, we know that in some types of projects there is some uncertainty as to just which activities will be included. In research projects in particular, several different outcomes of the project may result, depending upon the outcomes of certain chains of activities. A network that shows only one plan with one possible outcome may not represent adequately the true nature of the project.

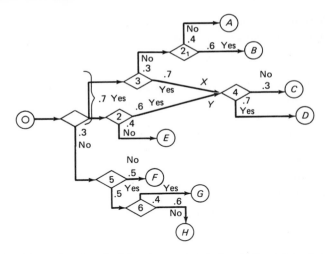

Figure 6-13   Example of a decision box network (excluding normal events).

Eisner[4] first published a description of this deficiency in PERT and CPM networks, and he formulated a network scheme that includes explicit treatment of alternative paths. In Eisner's scheme, events that lead to two or more alternative paths are called *decision boxes* (*dbs*), and a network containing such conditions is called a "*db* network."

As seen in Figure 6-13, a *db* is given a diamond shape with the alternative paths bursting from it. Each path emanating from a *db* is assigned a probability index, such that all paths derived from the same *db* will

Table 6-1.   Generalized Node Symbols Suggested by Pritsker and Happ[6]

Output ＼ Input	*Exclusive-or* ◁|	*Inclusive-or* ◁	*and* ◖
Deterministic, D	◁)	◁⟩	◯
Probabilistic, ▷	◁◇	◇	◯

    Exclusive-or—The realization of any branch leading into the node causes the node to be realized; however, one and only one of the branches leading into this node can be realized at a given time.

    Inclusive-or—The realization of any branch leading into the node causes the node to be realized. The time of realization is the smallest of the completion times of the activities leading into the *Inclusive-or* node.

    and—The node will be realized only if all the branches leading into the node are realized. The time of realization is the largest of the completion times of the activities leading into the *and* node.

    Deterministic—All branches emanating from the node are taken if the node is realized, that is, all branches emanating from this node have a *p*-parameter equal to one.

    Probabilistic—At most one branch emanating from the node is taken if the node is realized.

have probabilities that sum to one. These probabilities are estimated by he project planners in much the same way that activity durations are estimated—on the basis of judgment and experience with similar circumstances.

In a *db* network not all the activities will be performed, nor will all the possible project results be achieved. Therefore, it is of interest to the project planners to determine which outcomes or combinations of outcomes are possible, and then determine the probabilities of each possible outcome. A further step in the analysis would be to rank the outcomes according to their respective likelihoods. (Eisner's original procedures involve symbolic logic and conditional probability rules which will not be covered here.)

Eisner pointed out that when there is only one alternative, with probability of one, for each and every decision box in a network, then there is only one possible outcome of the project, also having a probability of one. Thus, the *db* network "degenerates" into a normal CPM or PERT network, indicating that the CPM-PERT concept is only a special case of the more general decision box approach.

Other scholars, especially Elmaghraby[5] and Pritsker [6, 7], have developed the generalized network idea further. They suggest a more comprehensive set of event-logic symbols, as shown in Table 6-1, to account for the various conditions that can be defined in a generalized or "stochastic" network. An illustration of these symbols used in the "inclusive-or" node situation is given in Figure 6-14.

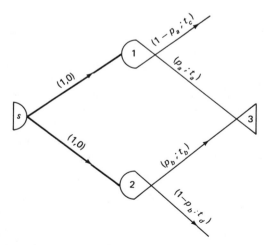

Figure 6-14 Sample of the Pritsker-Happ notation in a simple network with and *Inclusive or* Node.[6]

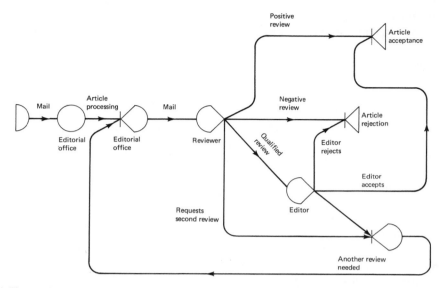

Figure 6-15 Generalized network model of a sequential review process.

Another application of this notation is given in Figure 6-15, which treats the process of reviewing an article submitted for publication in a professional journal. The network shows the possible relationships of reviewers and the editor of the journal, taking into account the possibilities that a reviewer may reject the article, and that the editor may accept or reject a qualified review or may require another review. Note that the network in addition to having branching nodes, also contains loops which are prohibited in the fully deterministic CPM-PERT networks.

Along with the more extensive graphical treatment of the generalized network concept, Elmaghraby, Pritsker, and others, have developed computational procedures that accommodate and exploit the more complex logic and stochastic nature of the generalized networks. The principal results of the computations are (1) the probability that a terminal node will be realized, and (2) the mean and variance of the duration of the project for each of the terminal nodes of interest.

Going beyond the usual "project" applications, Pritsker and others have shown how the stochastic network concept can be employed to formulate and solve certain problems in the fields of stochastic processes, including queuing models, inventory control, and reliability.

Although researchers in the field of generalized networks have written some computer programs to aid in certain calculations, there are no "packaged" programs for these procedures made available by the computer manufacturers, as is the case for normal CPM and PERT network calculations.*

As of this writing, generalized network theory is still under development, with only a few practical applications on record. It appears, however, that the concept offers at least an enlightening perspective on the scope of network theory, and will undoubtedly lead to practical solution methods for problems that have hitherto proved intractable. Two areas that appear to have considerable potential are in the development of information systems and in the planning and scheduling of expensive experimental programs.

Exercise 5 at the end of this chapter deals with the problem of drawing a GERT chart for the established policy of operation for a university patent and copyright committee. Another application of this type familiar to the authors is in the documentation, prior to computer programming of an inventory system dealing with large and expensive items. The network of the inventory policy has recycle loops to cover situations where requisitions are returned because of "out-of-control" unit prices. Branching nodes occur at points where material is committed from inventory to a particular project. The branches might be not enough lead time available, and not enough material available, in addition to the desired branch of committing the requested material.

There are several potential advantages of GERT networks in the above applications. First, the GERT network is quite useful in showing redundancies, inefficiencies, and inconsistencies in the policies or procedures in question. In executing policy, the GERT network could be quite useful in designing the information flow procedures. It could also be quite useful in speeding up action at each branching node in the network, since the requirements and implications of each alternative decision are quite clear.

Another intriguing potential application of GERT is in the planning and scheduling of the activities making up experiments that are associated with extremely expensive programs carried out in the hostile environments of space or on the ocean floor. These experiments have recycle loops and branching nodes. For example, in a space photographic experiment, cloud cover over the target area might require recycling to this activity at some later time. In a physiological experiment, the effects of certain activities on the astronauts may require proceeding with the experiment along one of several different branches. These experiments obviously require a high degree of preplanning and possibly on-line

* It is possible to use IBM's General Purpose Simulator System (GPSS) to make computations on stochastic networks with branching and feedback loops.

rescheduling capability to maximize the return from the experimental personnel and facilities. The GERT network appears to be an excellent vehicle for carrying out these tasks.

REFERENCES

1. Fondahl, J. W., *A Noncomputer Approach to the Critical Path Method for the Construction Industry*, 2nd Ed., Department of Civil Engineering, Stanford University, Stanford, Calif., 1962.
2. Archibald, Russell D., and R. L. Villoria, *Network-Based Management Systems*, John Wiley & Sons, New York, 1967 (Appendix B).
3. Anonymous, "Project Control System/360 (360A-CP-06X) Program Description and Operations Manual," No. H20-0376-0, IBM Corporation, White Plains, New York, 1967.
4. Eisner, Howard, "A Generalized Network Approach to the Planning and Scheduling of a Research Program," *Operations Research*, Vol. 10, No. 1, January–February 1962.
5. Elmaghraby, S. E., "An Algebra for the Analysis of Generalized Activity Networks," *Management Science*, Vol. 10, No. 3, April 1964.
6. Pritsker, Alan B., and W. William Happ, "GERT: Graphical Evaluation and Review Techniques, Part I. Fundamentals," *Journal of Industrial Engineering*, Vol. 17, No. 5, May 1966.
7. Pritsker, Alan B., and Gary E. Whitehouse, "GERT" Graphical Evaluation and Review Techniques, Part II. Probabilistic and Industrial Engineering Applications," *Journal of Industrial Engineering*, Vol. 17, No. 6, June 1966.

EXERCISES

1. The Air Force once tried a network scheme that avoided the errors and misunderstandings of the event scheme. Essentially it was the same as the event scheme, except that *every* activity was given both a start and a finish event. All the connections between the finish of one activity and the start of the next activity were made by dummy arrows. Sketch the market survey network strictly according to this discipline, and comment on why you think the Air Force abandoned the idea. Is there anything to be learned from this exercise relative to the precedence diagramming scheme illustrated in Figure 6-11?

2. Construct a table containing a list of criteria that you feel could be used to measure the utility of networking schemes. On the other axis list all the networking and project charting schemes you are familiar with, including hybrid forms of bar charts, milestone charts, or others. Then assign a rank or other value score to each scheme and criterion and summarize the score of each scheme. Which scheme received the highest score?

3. Compare the results of your scoring in the exercise above with those of other students and, if practical, with the opinions of persons experienced

with a variety of networking schemes. As a result of these comparisons, comment on whether it is practical to evaluate networking schemes on a technical, objective basis.

4. The techniques of resource allocation (Chapter 8) are applicable to certain types of job-shop scheduling problems, especially where the sequence of jobs is fixed and the problems are reduced to questions of the loading of certain facilities or pools of resources and the avoidance of project (or job) delays. Discuss how generalized network concepts may provide solution methods to a wider class of job-scheduling problems.

5. Draw a GERT network for the following procedure used by a university patent committee.
   1. Inventor submits invention to patent committee.
   2. Above (1) initiates two concurrent activities:
      2.1 Technical or commercial reviews by experts; and
      2.2 Legal review of inventor's patent liability to contractors and university.
   3. Study of legal and technical reviews.
   4. The study in (3) leads to
      4.1 Favorable evaluation, or
      4.2 Unfavorable evaluation
   5. Favorable evaluation in 4.1 leads to
      5.1 Submission of invention to university patent attorney; or
      5.2 Submission of invention to outside patent corporation.
   6. Activity 5.1 leads to
      6.1 University marketing of patent; or
      6.2 University drops invention because of lack of patent protection.
   7. Activity 5.2 leads to
      7.1 Outside patent corporation rejects invention; or
      7.2 Outside patent corporation seeks patent.
   8. Activity 7.1 leads to
      8.1 Resubmission by university patent committee of patent to outside patent corporation for reconsideration; or
      8.2 University drops invention.
   9. Activity 7.2 leads to
      9.1 Patent denied so university drops invention; or
      9.2 Patent acquired so outside patent corporation markets patent.
   10. Unfavorable evaluation in 4.2 leads to
      10.1 Submission of invention to outside patent corporation; or
      10.2 University drops invention.

7
COMPUTER PROGRAMMING OF THE BASIC SCHEDULING COMPUTATIONS

Hand methods of performing the basic network scheduling computations were taken up in Chapter 4, where numerical entries were placed directly on the arrow diagram. This is by far the most efficient hand method of carrying out these computations. If, however, networks are to be updated frequently, or if involved questions pertaining to resource allocation or time-cost trade-offs are raised, then the basic scheduling computations must be carried out many times on modified input data. For example, in the resource allocation procedures described in Chapter 8, the basic scheduling computations may be repeated several hundred times in the process of allocating resources to the project activities. In such cases, it is clear that the efficient computer processing of networks is very important. In Chapter 5 the problems associated with the use of computers were considered from a user's point of view. This chapter will deal with computer processing from the programmer's point of view.

Computer logic dictates that the basic scheduling computations must be carried out sequentially in some sort of tabular or matrix form. Thus, to introduce the subject of programming a computer to carry out these computations, tabular methods of hand computation will be described. This will be followed by the presentation of a computer flow diagram and a FORTRAN-IV program to carry out the basic scheduling computations for arrow diagrams. Finally, computer methods to carry out the basic scheduling computations will be taken up for the node diagramming method described in Chapter 6.

TOPOLOGICAL ORDERING OF PROJECT ACTIVITIES

The basic scheduling computational problem can be stated mathematically as follows, which is more brief and more suggestive than the form used in Chapter 4.

Earliest and Latest Event Times

$E_1 = 0$ by assumption, then

$$E_j = \max_i \ (E_i + D_{ij}) \quad ; \quad 2 \leqq j \leqq n. \tag{1}$$

$E_n =$ (Expected) project duration, and
$L_n = E_n$ or T_S, the scheduled project completion time. Then,

$$L_i = \min_j \ (L_j - D_{ij}) \quad ; \quad 1 \leqq i \leqq n - 1 \tag{2}$$

Earliest and Latest Activity Start and Finish Times and Slack

$$ES_{ij} = E_i \qquad ; \qquad \text{all } ij \tag{3}$$
$$EF_{ij} = E_i + D_{ij} \qquad ; \qquad \text{all } ij \tag{4}$$
$$LF_{ij} = L_j \qquad ; \qquad \text{all } ij \tag{5}$$
$$LS_{ij} = L_j - D_{ij} \qquad ; \qquad \text{all } ij \tag{6}$$
$$S_{ij} = L_j - EF_{ij} \qquad ; \qquad \text{all } ij \tag{7}$$

The above equations represent the basic scheduling computations previously taken up in Chapter 4. However, the emphasis here is on getting the earliest and latest event times, E_i and L_i, since the earliest and latest activity start and finish times follow directly from these in a straightforward fashion. If the computations are to be carried out in a progressive tabular form, then equation (1), for example, requires that the early start times (E_i's) for all activities merging to event j must have been computed prior to the computation of E_j. To insure this, the first step in most tabular procedures is to "arrange" the activities in a table or matrix so that the predecessors of any activity will always be found "above" it, or previous to it in the table, and successors will always be found "below," or after it. A table so arranged is referred to as *topologically ordered.* A simple way to accomplish this ordering is to number the network events so that $i < j$ for all activities, and then list the activities according to increasing i or j numbers. Fulkerson[1] has given the following simple procedure to accomplish this end:

(1) Number the initial project event (event with no predecessor activities) with 1. (If there is more than one initial project event, they should be numbered consecutively in any order.)

(2) Delete all activities from the initial event(s) and search for events in the new network that are now initial events; number these 2, 3, . . . , k in any order.

(3) Repeat step 2 until the terminal project event(s) is numbered.

The network originally given in Figure 4-5 is reproduced in Figure 7-1; it has its events numbered in topological order. The information

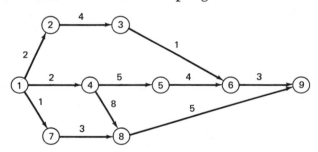

Initial Event i	Final Event j									Earliest Event Time, E_i
	1	2	3	4	5	6	7	8	9	
1	—	2		2			1			0
2	—	—	4							2
3	—	—	—			1				6
4	—	—	—	—	5			8		2
5	—	—	—	—	—	4				7
6	—	—	—	—	—	—			3	11
7	—	—	—	—	—	—	—	3		1
8	—	—	—	—	—	—	—	—	5	10
9	—	—	—	—	—	—	—	—	—	15
Latest Event Time, L_j	0	7	11	2	8	12	7	10	15	—

Figure 7-1 Arrow diagram and matrix representation of a project network

describing this network is also given in topological ordered matrix form. Each entry in the matrix corresponds to an activity in Figure 7-1. For example, in the row labeled initial (predecessor) event 1, we see entries in columns 2, 4, and 7. These are the final (successor) events of the three activities bursting from event 1. The entries of 2, 2, and 1 give the estimated duration times of their respective activities, 1-2, 1-4, and 1-7. Dashes have been placed in the lower left hand portion of this matrix, starting with the diagonal cells, to denote that activities corresponding to these cells are not possible in a topologically ordered matrix. Blanks in the upper right hand portion of this matrix indicate that while the corresponding activity is possible, it is not present in the particular network represented by the matrix.

We begin the computations by using the matrix in Figure 7-1 to carry out the computations of the earliest and latest event times given by equation (1) and (2). We first enter $E_1 = 0$ in the first row of the column giving the E_i's. Next, to find E_2, we note the entries in the column headed by $j = 2$. There is only one entry because only one activity in the network precedes event 2. We add this entry, $D_{1,2} = 2$, to the entry in the last column of this same row, $E_1 = 0$, to obtain $E_2 = 2$. The latter is entered in the second row of the E_i column. We proceed sequentially in this manner until we reach column 6, which is the first one to have more than one entry. In this case $E_3 + D_{3,6} = 7$ and $E_5 + D_{5,6} = 11$, so $E_6 = 11$; that is,

$$E_6 = \underset{i=3,5}{\text{Max}} \ (E_3 + D_{3,6} = 7, \ E_5 + D_{5,6} = 11) = 11.$$

To find the values of L_j, we start by arbitrarily letting $L_9 = E_9 = 15$, that is, the zero slack convention described in Chapter 4. For event 8, we note the entries along row 8. There is only one so that $L_8 = L_9 - D_{8,9} = 15 - 5 = 10$, which is entered in the L_j row for column 8. Again, we proceed sequentially in this manner until we reach row 4, which is the first row that arises with more than one entry. Here we note $D_{4,5} = 5$ and $D_{4,8} = 8$, so that L_4 becomes

$$L_4 = \underset{j=5,\,8}{\text{Min}} \ (L_5 - D_{4,5} = 3, \ L_8 - D_{4,8} = 2) = 2$$

Finally, we note that $L_1 = E_1 = 0$, which is a check on the accuracy of our computations, since we let $L_9 = E_9 = 15$.

The earliest and latest activity start and finish times and the total activity slack can now be computed in a straightforward manner using equations (3) through (7). These results have been given previously in Table 4-1 (with slightly different event numbers) and hence will not be repeated here. The sequential computational procedure illustrated in Figure 7-1 readily admits itself to computer programming.

COMPUTER PROGRAMMING

Programming the basic scheduling computations for computer processing is an exceedingly interesting exercise, because it permits ingenuity to be exercised to a great degree. Rather than following directly the computational procedure illustrated in Figure 7-1, a more general approach will be taken here which has a minimum of requirements placed on the input network data. In particular, the assumption that $i < j$ for all $i - j$, and the requirement of a unique pair of i and j numbers for each activity will not be made. The network must, of course, be free of loops and hence no duplication of event numbers is allowed. The ability to detect and to indicate the presence of such a network defect must, of course, be incorporated in the computational procedure to terminate the processing of such networks.

The logic for such a computational procedure is given in Figure 7-2. The presence of a loop in the network is detected by noting whether an activity has been added to the solution set in the forward pass computations, after each pass through the list of network activities. One should also note that the backward pass computations are greatly facilitated by preserving the order of processing the activities in making the forward pass computations, and then processing the activities in the reverse order in the backward pass. The computer code for this logic, written in FORTRAN-IV, is given in Figure 7-3 together with explanatory comment statements. There are 43 actual lines in this program.

The above program is just part of a much larger program which includes cost capabilities and network updating capabilities, including a graphical analysis of the actual activity performance times and costs, compared to the estimated quantities. This program has been applied to the network given in Figure 12-5 of Chapter 12, which contains 34 activities.

This program permits a wide variety of sorts of the basic output data. The output shown in Figure 7-4 is the result of a major sort on Total Slack (indicated by ****) and a minor sort on Early Start Date (indicated by ----). Thus, the first set of activities, having a slack of 14 working days, form the network critical path. Across the page for each activity are given the $i - j$ event number, followed by two columns for coding the activities referred to as RES, for resource type, and ID, for a second identification or basis of coding. In the illustrative example, resources 1, 2, 3, and 4 refer to design draftsmen, welders, mechanics, and procurement personnel, respectively. There is also interest in sum-

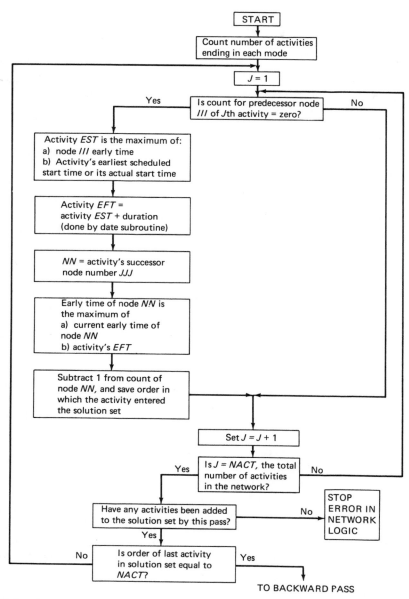

Figure 7-2 Flow diagram of logic for the program given in Figure 7-3.

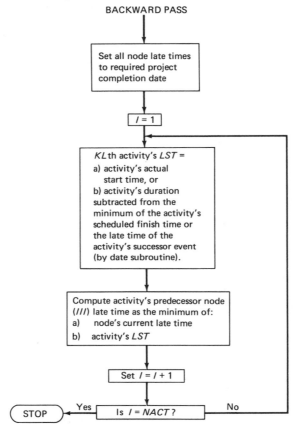

BACKWARD PASS

Set all node late times
to required project
completion date

$I = 1$

KL th activity's *LST* =
a) activity's actual
start time, or
b) activity's duration
subtracted from the
minimum of the activity's
scheduled finish time or
the late time of the
activity's successor event
(by date subroutine).

Compute activity's predecessor node
(*III*) late time as the minimum of:
a)   node's current late time
b)   activity's *LST*

Set $I = I + 1$

Is $I = NACT$?

Yes   STOP   No

Figure 7-2   Continued.

marizing the activities involving construction work; i.e., resource types
1, 2, and 3. Thus, these resources are assigned an ID number of 1 while
resource 4 is assigned an ID number of 2.

The column after the activity description gives the early start date,
which differs from the late start date in working days by an amount
equal to the total slack. The late start date, in turn, differs from the late
end date in working days by an amount equal to the activity duration.
The last column, giving dollar value complete, is merely the product
of the previous two columns.

This program exhibits a very interesting phenomenon that can only
occur in programs that are capable of treating activity time estimates

```
C
C
C        C.P.M. PROGRAM CODED FOR UNIVERSITY OF MIAMI 7040 COMPUTER
C        BY -- PABLO LARREA   9010 MILLER RD.  MIAMI, FLA.   33165
C
C
C        DESCRIPTION OF CONTENTS OF ARRAYS
C
C        NAME    DESCRIPTION
C        III     ACTIVITY I,J NODES (AS IIIJJJ)
C        DUR     ACTIVITY DURATION (TIME)
C        DESC    DESCRIPTION OF ACTIVITY
C        EST     ACTIVITY EARLY START TIME
C        LST     ACTIVITY LATE START TIME
C        JCOUNT  COUNT OF NO. OF ACTIVITIES INTO A NODE (EVENT)
C        IPOINT  ARRAY TO POINT TO ACTIVITY (USED TO SPEED UP SORTS)
C        LTIME   EVENT E.S.T. (FORWARD PASS) OR L.S.T. (BACKWARD PASS)
C        IORDER  ORDER IN WHICH ACTIVITIES WERE ENTERED IN FORWARD PASS
C                  (BACKWARD PASS WILL ENTER THEM IN REVERSE ORDER)
C        ISTRT   ACTIVITY EARLIEST START TIME POSSIBLE (BY DATA)
C        IEND    ACTIVITY LATEST FINISH TIME POSSIBLE (DUE DATE BY DATA)
C
C
C        INPUT PROCEDURE HERE
C
C
C        BEGIN THE C.P.M. PROCEDURE
C        COUNT ACTIVITIES TO A NODE
         DO 520 I=1,NACT
         KL=MOD(III(I),1000)
     520 JCOUNT(KL)=JCOUNT(KL)+1
C        SET POINTER ARRAY TO POINT TO IIIJJJ
         DO 570 I=1, NACT
     570 IPOINT(I)=III(I)*2000+I
         K=0
         LS=NACT
     580 L=LS
         LS=0
         DO 620 J=1,L
         I=IPOINT(J)/2000000
C        NOW LOOK FOR THOSE ACTIVITIES THAT BEGIN IN A NODE WHOSE COUNT
C        HAS BECOME 0
         IF (JCOUNT(I).EQ.0)  GO TO 600
C        CRITERIA NOT SATISFIED --- MOVE ITS POINTER UP FOR NEXT TIME ---
         LS=LS+1
         IPOINT(LS)=IPOINT(J)
         GO TO 620
C        NOW WE HAVE ONE WITH ALL PREREQUISITES SATISFIED
C        INDICATE WE HAVE FOUND SOME
     600 KL=MOD(IPOINT(J),2000)
C        MARK ACTIVITY E.S.T. AS TIME FOR NODE PRECEEDING IT
         EST(KL)=MAXO(LTIME(I),ISTRT(KL))
         K=K+1
```

Figure 7-3   FORTRAN IV program for the logic shown in Figure 7-2 dealing with the basic scheduling computations for arrow diagrams.

```
C     SAVE FORWARD PASS SEQUENCE TO SPEED UP BACKWARD PASS
      IORDER(K)=KL
      NN=MOD(III(KL),1000)
C     CALCULATE NOW FOR SUCCEEDING NODE
      CALL DATE(3,KL)
      LTIME(NN)=MAXO(LTIME(NN),KL)
C     DECREMENT COUNT OF PREREQUISITES INTO JJJ NODE
      JCOUNT(NN)=JCOUNT(NN)-1
  620 CONTINUE
      IF(K.EQ.NACT) GO TO 625
      IF(LS.NE.0) GO TO 580
C     ERROR IN LOOPING NOTIFY USER
      WRITE (6,622) LS
  622 FORMAT (1H1,10X,27HLOOP ENCOUNTERED CHECK DATA    ,I4,
     1 22H REMAINING ACTIVITIES   )
      WRITE(6,624) (IPOINT(I),I=1,LS)
  624 FORMAT (10I13)
      CALL EXIT
C     NOW TO PREPARE FOR THE BACKWARD PASS
  625 DO 640 I=1,NACT
      KL=MOD(III(I),1000)
  640 LTIME(KL)=IFINAL
      IT=LTIME(NN)
C     NOW FOR BACKWARD PASS
      DO 650 I=1,NACT
C     GET REVERSE ORDER FROM FORWARD PASS
      KL=NACT-I+1
      KL=IORDER(KL)
      NN=MOD(III(KL),1000)
C     GET ACTIVITY L.S.T.
      CALL DATE(4,KL)
      NN=III(KL)/1000
C     COMPUTE L.S.T. FOR III NODE
  650 LTIME(NN)=MINO(LTIME(NN),MOD(LST(KL),10000))
C     NOTE NOW WE HAVE EST AND LST ARRAYS ONLY
C     TO COMPUTE SLACK, LFT,ETC APPROPRIATE  LOOPS SHOULD BE WRITTEN
C
C     OUTPUT PROCEDURE HERE
```

Figure 7-3 Continued.

in either working days or calendar days. The latter is marked by a (C) after the duration time. The phenomenon referred to is the fact that activities 1-5, 5-6, and 6-7 each have 15 days of slack, whereas the remainder of this critical path, i.e., 7-8, 8-20, 20-21, etc., has only 14 days of slack. The circumstances that give rise to this peculiar result is that activity 6-7, whose duration is 14 *calendar* days, encompasses one holiday on the forward pass, i.e., Thanksgiving Day, 11/28/68. However, on the backward pass, the 14 calendar days for this activity encompasses two holidays, i.e., 12/24/68 and 12/25/68, which were both taken as holidays in this project. As a result, this activity, and those preceding it have one more day of slack than the remainder of the critical path.

```
        SAMPLE PROBLEM -- CONSTRUCTION OF AN OCEANOGRAPHIC CABLE ENGINE                    PAGE    3
        DATE OF REPORT 10/30/68          PROJECT START DATE 10/31/68        REQUIRED COMPLETION DATE  1/31/69
```

I NODE	J NODE	RES	ID	ACTIVITY DESCRIPTION	EARLY ST. DATE	TOTAL SLACK	LATE ST. DATE	DURA-TION	LATE END DATE	PERCENT ELAPSED	TOTAL $ VALUE	$ VALUE COMPLETE
7	8	2	1	FABRICATE DRUM	12/ 4/68	14	12/26/68	10	1/10/69		3000	
8	20	2	1	INSTALL HUB/DRUM	12/18/68	14	1/10/69	5	1/17/69		2250	
20	21	2	1	INSTALL BRAKE/KNIFE	12/27/68	14	1/17/69	2	1/21/69		300	
21	23	3	1	SUBPLATE/GEARS/HYDRS	12/31/68	14	1/21/69	5	1/28/69		750	
21	22	2	1	SUBPLATE/GEARS/HYDRS	12/31/68	14	1/21/69	5	1/28/69		1500	
23	22			DUMMY	1/ 8/69	14	1/28/69	0C	1/28/69		-0	
22	24	3	1	TEST ASSEMBLY	1/ 8/69	14	1/28/69	2	1/30/69		600	
24	25	3	1	INSTALL IN SHIP	1/10/69	14	1/30/69	1	1/31/69		150	
1	5	1	1	DRUM DRAWINGS	10/31/68	15	11/22/68	10	12/ 9/68		1250	
5	6	4	2	ORDER HUB COMPONENTS	11/15/68	15	12/ 9/68	3	12/12/68		-0	
6	7	4	2	REC. HUB COMPONENTS	11/20/68	15	12/12/68	14C	12/26/68		500	
1	16	1	1	HYDRAULIC DETAILS	10/31/68	18	11/27/68	5	12/ 5/68		625	
1	18	1	1	GEAR BOX SPECS.	10/31/68	18	11/27/68	5	12/ 5/68		625	
16	17	4	2	HYDRAULIC COMPONENTS	11/ 8/68	18	12/ 5/68	1	12/ 6/68		25	
18	19	4	2	ORDER GEAR BOX	11/ 8/68	18	12/ 5/68	1	12/ 6/68		50	
17	21	4	2	RECEIVE HYDRAULICS	11/11/68	18	12/ 6/68	45C	1/20/69		1750	
19	21	4	2	RECEIVE GEAR BOX	11/11/68	18	12/ 6/68	45C	1/20/69		1250	
1	11	1	1	BRAKE DRAWINGS	10/31/68	20	12/ 2/68	15	12/23/68		1875	
11	12	4	2	ORDER BRAKE MATL'S	11/22/68	20	12/23/68	2	12/27/68		50	
12	13	4	2	REC BRAKE MATERIALS	11/26/68	20	12/27/68	14C	1/10/69		450	
13	20	2	1	FABRICATE BRAKE	12/10/68	20	1/10/69	5	1/17/69		1500	
5	7	1	1	DRUM WORKLIST	11/15/68	21	12/17/68	5	12/26/68		300	
4	8	2	1	FABRICATE FOUNDATION	11/18/68	25	12/26/68	10	1/10/69		3000	
1	2	1	1	PREPARE FOUNDATION	10/31/68	27	12/11/68	5	12/18/68		625	
2	4	4	2	FOUNDATION WORKLIST	11/ 8/68	27	12/18/68	4	12/26/68		500	
2	3	4	2	ORDER FOUND. MAT'S	11/ 8/68	28	12/19/68	1	12/20/68		50	
3	4	4	2	REC. FOUND. MAT'S	11/11/68	28	12/20/68	6C	12/26/68		1000	
11	13	4	2	BRAKE WORKLIST	11/22/68	29	1/ 8/69	2	1/10/69		250	
1	9	1	1	KNIFE DRAWINGS	10/31/68	31	12/17/68	12	1/ 7/69		1500	
9	10	4	2	KNIFE WORKLIST	11/19/68	31	1/ 7/69	3	1/10/69		200	
10	20	2	1	FABRICATE KNIFE	11/22/68	31	1/10/69	5	1/17/69		1250	
1	14	1	1	SUBPLATE SPECS.	10/31/68	35	12/23/68	2	12/27/68		250	
14	15	4	2	ORDER SUBPLATE	11/ 4/68	35	12/27/68	2	12/31/68		50	
15	21	4	2	RECEIVE SUBPLATE	11/ 7/68	35	12/31/68	21C	1/21/69		250	

```
        ***TOTAL***                                                                        27725
```

Figure 7-4 Basic computer output-initial run, with total slack major sort and early start date as secondary sort.

The project start date is 10/31/68 and the due date is 1/31/69. An update of this project, as of 11/15/68, is given in Figure 7-5. Activities with no late start dates have begun, and hence their early start dates are actual start dates. Also, the late end dates are actual end dates where the percent elapsed is 100. In this particular update, we see that the total slack has slipped from 14 to 12 working days, and $3494 of the estimated total of $27,725 has been expended.

Special Sort Reports

Outputs sorted by resource types are shown in Figure 7-6. The output for resource 1, which has a minor sort by early start date, is given at the top of this figure. Similar outputs are produced for resources 2, 3,

SAMPLE PROBLEM -- CONSTRUCTION OF AN OCEANOGRAPHIC CABLE ENGINE   PAGE   3
DATE OF REPORT 11/15/68   PROJECT START DATE 10/31/68   REQUIRED COMPLETION DATE 1/31/69

I NODE	J NODE	RES	ID	ACTIVITY DESCRIPTION	EARLY ST. DATE	TOTAL SLACK ****	LATE ST. DATE	DURA-TION	LATE END DATE	PERCENT ELAPSED	TOTAL $ VALUE	$ VALUE COMPLETE
5	6	4	2	ORDER HUB COMPONENTS	11/15/68			3	11/20/68	33.3	-0	-0
19	21	4	2	RECEIVE GEAR BOX	11/14/68			45C	12/30/68	4.4	1250	55
1	2	1	1	PREPARE FOUNDATION	11/14/68			5	11/21/68	40.0	625	249
18	19	4	2	ORDER GEAR BOX	11/13/68			1	11/14/68	100.0	50	50
17	21	4	2	RECEIVE HYDRAULICS	11/13/68			45C	12/30/68	6.7	1750	116
1	11	1	1	BRAKE DRAWINGS	11/12/68			15	12/ 4/68	26.7	1875	499
16	17	4	2	HYDRAULIC COMPONENTS	11/11/68			1	11/12/68	100.0	25	25
1	18	1	1	GEAR BOX SPECS.	11/ 6/68			5	11/13/68	100.0	625	625
1	16	1	1	HYDRAULIC DETAILS	11/ 1/68			5	11/11/68	100.0	625	625
1	5	1	1	DRUM DRAWINGS	10/30/68			10	11/14/68	100.0	1250	1250
13	20	2	1	FABRICATE BRAKE	12/20/68	12	1/10/69	5	1/17/69		1500	
20	21	2	1	INSTALL BRAKE/KNIFE	12/31/68	12	1/17/69	2	1/21/69		300	
21	23	3	1	SUBPLATE/GEARS/HYDRS	1/ 3/69	12	1/21/69	5	1/28/69		750	
21	22	2	1	SUBPLATE/GEARS/HYDRS	1/ 3/69	12	1/21/69	5	1/28/69		1500	
23	22			DUMMY	1/10/69	12	1/28/69	0C	1/28/69		-0	
22	24	3	1	TEST ASSEMBLY	1/10/69	12	1/28/69	2	1/30/69		600	
24	25	3	1	INSTALL IN SHIP	1/14/69	12	1/30/69	1	1/31/69		150	
11	12	4	2	ORDER BRAKE MATL'S	12/ 4/68	13	12/23/68	2	12/27/68		50	
12	13	4	2	REC BRAKE MATERIALS	12/ 6/68	13	12/27/68	14C	1/10/69		450	
7	8	2	1	FABRICATE DRUM	12/ 4/68	14	12/26/68	10	1/10/69		3000	
8	20	2	1	INSTALL HUB/DRUM	12/18/68	14	1/10/69	5	1/17/69		2250	
6	7	4	2	REC. HUB COMPONENTS	11/20/68	15	12/12/68	14C	12/26/68		500	
4	8	2	1	FABRICATE FOUNDATION	11/29/68	17	12/26/68	10	1/10/69		3000	
2	4	4	2	FOUNDATION WORKLIST	11/21/68	18	12/18/68	4	12/26/68		500	
2	3	4	2	ORDER FOUND. MAT'S	11/21/68	19	12/19/68	1	12/20/68		50	
3	4	4	2	REC. FOUND. MAT'S	11/22/68	19	12/20/68	6C	12/26/68		1000	
5	7	1	1	DRUM WORKLIST	11/18/68	20	12/17/68	5	12/26/68		300	
1	9	1	1	KNIFE DRAWINGS	11/18/68	20	12/17/68	12	1/ 7/69		1500	
9	10	4	2	KNIFE WORKLIST	12/ 5/68	20	1/ 7/69	3	1/10/69		200	
10	20	2	1	FABRICATE KNIFE	12/10/68	20	1/10/69	5	1/17/69		1250	
11	13	4	2	BRAKE WORKLIST	12/ 4/68	22	1/ 8/69	2	1/10/69		250	
1	14	1	1	SUBPLATE SPECS.	11/18/68	24	12/23/68	2	12/27/68		250	
14	15	4	2	ORDER SUBPLATE	11/20/68	24	12/27/68	2	12/31/68		50	
15	21	4	2	RECEIVE SUBPLATE	11/22/68	24	12/31/68	21C	1/21/69		250	

TOTAL   27725   3494

Figure 7-5   Update of network whose initial computer run is shown in Figure 7-4

and 4, and also for ID numbers of 1 and 2. At the middle and bottom of Figure 7-6 are summary reports based on RESource numbers and ID numbers. The computer program also has the capability of presenting this update time and cost performance information in graphical form. For each RESource type the ratio of actual working days (or costs) to estimated working days (or costs) for all activities completed are plotted against the number of elapsed working time units. Such graphs (not shown in this text) are evidently preferred by some management personnel over tabular forms of presenting update information.

NODE DIAGRAMMING COMPUTATIONAL PROCEDURES

The node method of networking was described in detail in Chapter 6. It was noted that although this system of networking is not widely used today, it has many natural advantages, and its use is expected to increase in the future. The problems of drawing networks and making the basic scheduling computations by hand were illustrated in Figures 6-4 and 6-5.

I NODE	J NODE	RES **	ID	ACTIVITY DESCRIPTION	EARLY ST. DATE	TOTAL SLACK	LATE ST. DATE	DURA-TION	LATE END DATE	PERCENT ELAPSED	TOTAL $ VALUE	$ VALUE COMPLETE
1	5	1	1	DRUM DRAWINGS	10/30/68			10	11/14/68	100.0	1250	1250
1	16	1	1	HYDRAULIC DETAILS	11/ 1/68			5	11/11/68	100.0	625	625
1	18	1	1	GEAR BOX SPECS.	11/ 6/68			5	11/13/68	100.0	625	625
1	11	1	1	BRAKE DRAWINGS	11/12/68			15	12/ 4/68	26.7	1875	499
1	2	1	1	PREPARE FOUNDATION	11/14/68			5	11/21/68	40.0	625	249
5	7	1	1	DRUM WORKLIST	11/18/68	20	12/17/68	5	12/26/68		300	
1	9	1	1	KNIFE DRAWINGS	11/18/68	20	12/17/68	12	1/ 7/69		1500	
1	14	1	1	SUBPLATE SPECS.	11/18/68	24	12/23/68	2	12/27/68		250	

```
                     ***TOTAL***                                                              7050       3248
```

RES	TOTAL $ VALUE	$ VALUE COMPLETE	RES	TOTAL $ VALUE	$ VALUE COMPLETE
1	7050	3248	2	12800	0
3	1500	0	4	6375	246

ID	TOTAL $ VALUE	$ VALUE COMPLETE	ID	TOTAL $ VALUE	$ VALUE COMPLETE
1	21350	3248	2	6375	246

Figure 7-6 Update computer output—initial run, with resource type major sort and early start date minor short (only resource 1 is shown); together with summary of dollar value for each resource type.

There are various ways to represent project arrow and node networks. A comparison of these is made by Dimsdale,[2] who also discusses how to transform the arrow representation to the node representation. Figure 7-7 illustrates the graphic and symbolic conventions customarily employed to describe node networks. Figure 7-7a displays a node network and an associated table which gives duration times for each activity. Figure 7-7b represents the same network by means of a precedence matrix, P, and a duration vector D. This network will be used to describe an extremely clever computational procedure developed by Montalbono,[3] which requires only a single pass through the network. This procedure, the steps which are illustrated in Figure 718, use the following conventions:

(1) A node with no arrows pointing at it, or with only dashed arrows pointing at it, will be called a *source*.

(2) A node with no arrows pointing away from it, or with only dashed arrows pointing away from it, will be called a *sink*.

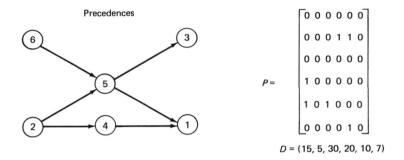

$$P = \begin{bmatrix} 0 & 0 & 0 & 0 & 0 & 0 \\ 0 & 0 & 0 & 1 & 1 & 0 \\ 0 & 0 & 0 & 0 & 0 & 0 \\ 1 & 0 & 0 & 0 & 0 & 0 \\ 1 & 0 & 1 & 0 & 0 & 0 \\ 0 & 0 & 0 & 0 & 1 & 0 \end{bmatrix}$$

$D = (15, 5, 30, 20, 10, 7)$

DURATIONS

Node	Time
1	15
2	5
3	30
4	20
5	10
6	7

(a)

Successors	Source activity	Duration	Predecessors
	1	15	4,5
4,5	2	5	—
—	3	30	5
1	4	20	2
1,3	5	10	2,6
5	6	7	—

(b)

Figure 7-7   Methods of presenting a node network.

(3) Those sources for which forward times have not been calculated at the start of an iteration will be called *new sources*. Those for which times *have* been calculated will be called *old sources*. Similarly for *backward times* and *new* and *old* sinks.

NARRATIVE DESCRIPTION OF THE BASIC ALGORITHM

The earliest time at which an activity can start is the time at which the latest of its predecessors is finished. The latest time at which an activity can finish is the time at which the earliest of its successors must start. This symmetry can be used to calculate backward times in a manner exactly analogous to calculating forward times and *concurrently with the calculation of forward times*, i.e., without a prior determination of the critical-path or minimum project duration time. Thus, steps (a), (b), and (c) of Figure 7-8 are accomplished in three scans of the precedence matrix, rather than the six which would be required if forward times were calculated before backward times. It is most important to keep this point in mind in considering the following discussion.

The steps in the basic algorithm are described below in terms of operations on nodes and arrows like those in Figure 7-8:

(1) Locate new sources and new sinks (see definition above). If there are none, the algorithm is at an end. The ending procedure is described in step 5. Note the parallel treatment of forward and backward calculations in Figure 7-8.

(2) Calculate forward times for new sources and backward times for new sinks. A *forward time* is calculated for each new source as the sum of its duration time plus the forward time of the one or more of its predecessors, if any, whose forward time is greatest. Note that predecessors are connected to new sources by dashed arrows pointing *at* the new source. A *backward time* is calculated for each new sink as the sum of its duration time plus the backward time of the one or more of its successors, if any, whose backward time is greatest. Note that successors are connected to new sinks by dashed arrows pointing *away* from the new sink.

(3) Replace, by dashed arrows, all the solid arrows pointing *away* from the new sources which have just been processed. Replace, by dashed arrows, all the solid arrows pointing *at* the new sinks which have just been processed. For consistent networks, this will either develop new sources and sinks for the next iteration, or, if no solid arrows remain, it will indicate that the calculations are complete.

(4) Repeat step 1.

(5) *Ending procedure.* If all the nodes in the network have both forward and backward times associated with them, the network is consistent. In this case, an *early start time* can be calculated for each node by subtracting its duration time from its forward time; a *late start time* can be calculated for each node by subtracting its backward time from the largest backward (or forward) time found in the network. This largest time is the critical-path or minimum project duration time. Those nodes whose early and late start times are equal lie on the critical path.

If the nodes do not all have both forward and backward times, the network is inconsistent, that is, it contains one or more loops.

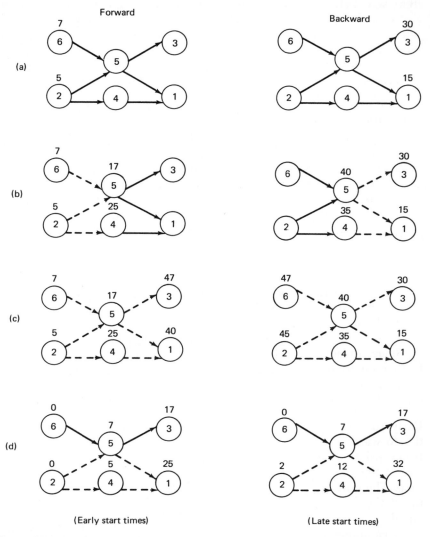

Figure 7-8 Graphic representation of steps in algorithm for consistent network.

In Figure 7-8, sections (a), (b) and (c) depict the three iterations required by the basic algorithm to calculate forward and backward times. Section (d) depicts the calculations of early and late start times after the basic algorithm is completed.

ITERATION (a). The new sources (in this case, the *original sources*) are nodes 2 and 6. The *new sinks* are nodes 1 and 3. Since the new sources have no predecessors and the new sinks have no successors, the forward and backward times are merely the duration times of each activity. These are written above their corresponding nodes, as are all the times calculated in subsequent iterations.

ITERATION (b). New sources: Nodes 4 and 5. New sinks: Nodes 4 and 5. Forward times: Node 4—predecessor time (5) plus duration time (2) = 25; Node 5—greatest predecessor time (7) plus duration time (10) = 17. Backward times: Node 4—successor time (15) plus duration time (2) = 35; Node 5—greatest successor time (30) plus duration time (10) = 40.

ITERATION (c). New sources: Nodes 1 and 3. New sinks: Nodes 2 and 6. Forward times: Node 1—greatest predecessor time (25) plus duration time (15) = 40; Node 3—predecessor time (17) plus duration time (30) = 47. Backward times: Node 2—greatest successor time (40) plus dura-time (5) = 45: Node 6—successor time (40) plus duration time (7) = 47.

Since no new sources or sinks are developed by iteration (c), no further iterations are needed. The process of calculating early and late start times ends with:
STEP (d).

Early start times are calculated by subtracting activity duration from the forward times. Late start times are calculated by subtracting the backward times from the project duration time, in this case, 47—the largest time associated with any node either as a forward or a backward time. The critical path, indicated by solid arrows in figure (d), is made up of those nodes (activities) for which early and late start times are equal.

A computer logic flow diagram for the above computational procedure is given in Figure 7-9. A computer program written in Iverson's APL language is given by Montalbano;[3] it contains a remarkable total of only 13 program lines. A FORTRAN program for this procedure will be left to the reader as an exercise.

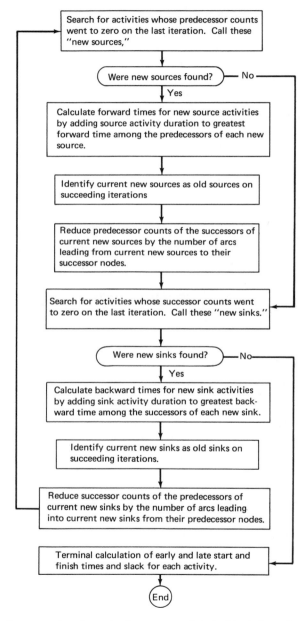

Figure 7-9 Programming flow chart corresponding to Montalbano's node network computational procedure.

REFERENCES

1. Fulkerson, D. R., "Expected Critical Path Lengths in PERT Networks," *Operations Research,* Vol. 10, No. 6, November–December (1962), pp. 808-817.
2. Dimsdale, B., "Computer Construction of Minimal Project Networks," *IBM Systems Journal,* Vol. 2, March 1963.
3. Montalbano, M., "High-Speed Calculation of the Critical Paths of Large Networks," IBM Systems Research and Development Center *Technical Report,* Palo Alto, California.

EXERCISES

1. Draw up a table similar to 7-1 for the network developed in exercise 5 of Chapter 4, or exercise 1 of Chapter 9, using only the normal activity duration times.
2. Draw a computer flow diagram representing the logic you would use to program the scheduling computation procedure described in Table 7-1.
3. Draw a node diagram for the network in Figure 7-1 and carry out the computational procedure developed by Montalbano and illustrated in Figure 7-8.
4. Write a computer program following the logic developed in exercise 2.
5. Write a computer program following the node diagram logic shown in Figure 7-7, to perform the basic scheduling computations.

8

SCHEDULING ACTIVITIES TO SATISFY RESOURCE CONSTRAINTS

In previous chapters we have developed the method of expressing a particular plan to carry out a project in the form of an arrow diagram. By adding time estimates to the activities we were then able to perform the basic scheduling computations, giving the early and late start and finish times and the slack times for each activity in the project. We are now at the point where we can determine how the slack should be allocated to various activities so as to alleviate excessive demands on key resources, such as personnel or physical resource items.

It would be advisable at this point for the reader to refer to Figure 1-4 which depicts the overall network-based planning and control procedure. This figure indicates how resource allocation (box 4) fits into the overall planning procedure. This chapter will consider how to satisfy resource constraints placed on a project, not only by judicious scheduling of activities on slack paths, but also by changing the duration times and resource loading of certain activities, by splitting (interrupting) certain activities, and by other devices commonly employed in the practice of carrying out project type work.

Resource allocation is probably receiving more attention today than any other aspect of PERT and CPM. There are several very good reasons for this. First, the significance of the problem is growing rapidly. Modern technology has developed many large and expensive physical resources which must be accounted for. More important, however, is the fact that the number of different personnel resources is increasing due to growing specialization and new technologies. Personnel resources come in different "trades" which are further broken down by skill, geographical location, departmental barriers, etc. For these reasons, it is not uncommon to deal with problems where 25 to 50 or more different resources must be considered.

Another reason for the current interest in resource allocation procedures is that the sheer magnitude of the problem is so great that central planning cannot explore many of the possible alternatives. Considering the scheduling possibilities cited above, such as scheduling activities on slack paths, changing activity duration times and resource loading, and splitting activities, we see that we have a combinatorial problem of a most formidable magnitude. In fact, the largest computers could not explore all of the possibilities in a reasonable length of time in anything but trivially small networks. Furthermore, the development of an optimal procedure does not appear likely in the foreseeable future since there is currently no mathematical basis for a *realistic* scheduling procedure. While the problem with suitable restrictive assumptions can be formulated as an integer linear programming problem, such as done in Appendix 9-1, there is little hope of solving the problem except for trivially small networks. Wiest[1] states this quite well by saying, ". . . the use of linear programming and a 7090 (computer) for such problems would be somewhat akin to using a bulldozer to move a pebble." Wiest further suggests the use of heuristic (rule of thumb) techniques, which have the ability to rapidly generate solutions for solving this problem. This ability, together with the use of cues in the problem environment to narrow our search to a subspace rich in good schedules, appears to be better than trying to exhaustively search the space of possible schedules for the best one.

In Chapter 5 the characteristics of a number of computer-oriented scheduling procedures were discussed. The last section of this chapter describes several such procedures for multiproject multiresource scheduling. It is not, however, the intent of this chapter to catalogue all of the current procedures, but rather to structure the more important problems, and to use them as illustrations of network-based scheduling procedures. It is hoped that these illustrations will furnish a basis for the reader to innovate new scheduling procedures for specific applications.

SIMPLE RESOURCE ALLOCATION CHECK

The simplest resource loading check might be described as noting, from a visual perusal of the network, the occurrence of demand for the resources in question. Assuming activities are scheduled to start at their earliest times (which are written on the network), one can note overlaps in the time periods when the resource is scheduled and thus observe when, if ever, the demand for the resource is excessive. This type of checking is facilitated by a time scaled network. But this is not necessary if the computations have been made on the network using the special symbols described in Chapter 4. If resource overlaps occur, alternate activity schedules can be tested, keeping within the bounds dictated by the respective activity slack figures which should also be recorded on the network.

As an example of this sort of resource allocation check, consider the

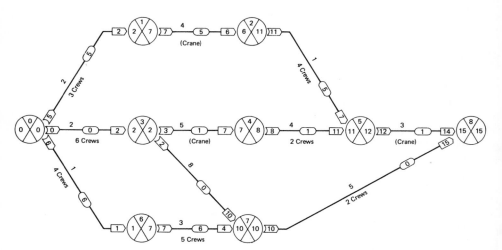

Figure 8-1 Project network with activity crew requirements listed.

network shown in Figure 8-1, which was originally presented in Figure 4-5. Suppose you are required to employ a large crane on activities 1-2, 3-4, and 5-8 on one activity at a time and during their entire duration. Will completion of the project be delayed? The network shown in Figure 8-1 permits you to answer "no." First, from the network we can see that technologically, activity 1-2 can be performed before or after activity 3-4. However, they both must be finished before activity 5-8 is started. Thus, the problem reduces to deciding which of the activities 1-2 and 3-4 should be scheduled first. Since they both have the same early start time, i.e., the end of time 2, it is logical to base the choice on slack and schedule activity 3-4 first. Thus, the following feasible schedule can be obtained quite easily by inspection.*

Activity	Scheduled Activity Start and Finish Times
3-4	2-7 (earliest possible schedule)
1-2	7-11 (latest possible schedule)
5-8	12-15 (latest possible schedule)

The above type of problem can be handled quite well by informal procedures. Consider, however, a more involved problem, such as one suggested by the crew requirements shown beneath the activities in Figure 8-1. If all activities are started as early as possible, a total of 13 crews would be required during the first time period, 14 during the second period, etc. Suppose only 8 crews are available for assignment to this project. Will this constraint require an increase in the duration of the project? If so, what is the minimum project duration that can be achieved? Suppose we can hire as many crews as necessary, but we would like to minimize the maximum number required at any one time. How many crews would this call for? Suppose management wants to know how long the project will take as a function of the number of crews assigned to the project. If this information is available, what is the optimum total crew assignment, considering labor costs as well as the "value" of completing the project in various lengths of time?

---

* This idea can sometimes be used to build certain restrictions into the network that cannot be expressed by network logic alone. For example, suppose there are several activities that cannot go on simultaneously for some reason, but otherwise they are independent of each other. For instance, the activities might be different maintenance jobs in the cockpit of an airplane which can physically accommodate only one maintenance crew at a time. By creating an artificial resource called "cockpit," we can insure that these two activities will not occur simultaneously by employing a scheduling procedure which allocates only one cockpit at a time. The opposite situation also occurs where two or more activities must be scheduled concurrently or not at all. This requirement can be satisfied by creating a threaded list of such activities programmed so that all or none of the activities in the list are scheduled.

These are very practical questions that arise in many different problem situations. We will now structure the resource allocation problem to indicate how these questions might be answered.

DESCRIPTION OF BASIC SCHEDULING PROBLEMS

To formulate specific answers to the above types of questions, the following basic assumptions will be made:

(1) The projects to be scheduled each have an assigned (perhaps tentative) start date and a due date.
(2) Each of the projects is characterized by a technological ordering of the project activities in the form of an arrow diagram.
(3) The resource levels available by time period are specified for each of the resource types being considered.
(4) The resource requirements of each activity are specified and are assumed to be constant during the duration of the activity.

The first two assumptions will be retained throughout this chapter. The last two assumptions will be subsequently relaxed to simulate a more realistic type of scheduling situation.

On the basis of the above assumptions, scheduling problems can be roughly classified into three groups which will be referred to as un-limited resource leveling, limited resource allocation, and long range resource planning.

1. *Unlimited Resource Leveling.* This problem arises when it is possible to procure sufficient resources to carry out a project which must be completed by a specified due date. This situation exists, for example, in many types of construction work. The scheduling objective in this case is to minimize the resource costs. Since the cost of hiring and laying off personnel or physical resources are appreciable, this objective is equivalent to *leveling*, as much as possible, the demand for each specific resource during the life of the contract, with perhaps an initial buildup period and a terminal tapering off period. Thus, the objective here is to *level the resource requirements, subject to the constraint that the project due date must be met.*

2. *Limited Resource Allocation.* This problem, which is the more common situation, arises when there are very definite limitations on the resources available to carry out the project or projects under considera-tion. In this case, the scheduling objective is to meet project due dates insofar as possible, which is equivalent to *minimizing the duration of*

the projects being scheduled, subject to stated constraints on available resources.

3. *Long Range Resource Planning.* This problem is a generalization of the Limited Resource Allocation problem. It arises in the context of long range resource planning, where management seeks to determine the combination of resource levels and project due dates that will minimize resource costs, overhead costs, and losses which result when project due dates are not met. If this resource planning is for a single project, then the procedure given here can also be referred to as a solution to the time-cost trade-off problem which is taken up formally in the next chapter. The long range resource planning problem is by far the most difficult of the three basic scheduling problems cited here, because the problem has the fewest constraints.

The basic approach to be followed in solving each of these problems is to first order the activities according to some criterion, and then to schedule the activities in the order listed as soon as their predecessors are completed and adequate resources are available.

Davis[2, 3] has published a survey of procedures of this type, as well as a procedure of his own which will give optimal solutions to modest sized networks under the restricted set of assumptions given above. Some of the more important papers in the development of this subject have been written by Clark[4] (1961), Burgess[5] (1962), McGee[6] (1962), Kelley[7] (1963), Moshman[8] (1963), Mize[9] (1964), Conway[10] (1964), Brooks[11] (1964), Wiest[12] (1967), and Fendley[13] (1967).

The key element in the basic scheduling approach suggested above is the criterion used to order the activities for scheduling. A change in the order will, of course, change the resulting schedule. In almost all studies of this problem the intuitively reasonable criterion of least slack first is recommended. There is considerable empirical evidence that this criterion generally gives less project slippage and idle resources than other criteria. It will not always, of course, give the best schedule. In fact, a network can always be contrived that will favor one ordering criterion over another.

There is also some evidence supporting the secondary ordering criterion of shortest activity duration first. If two activities with the same slack are eligible for scheduling, the one with the shortest duration should be scheduled first. This criterion has the obvious appeal that if only one of two activities can be scheduled at a time, the wait time is minimized by scheduling the shorter activity first. There is hardly any purpose of considering a tertiary ordering criterion.

Of the three problems cited above, the case of limited resource allocation will be considered first. The reason for this is merely that it will later serve as the basis for solving the other two problems.

LIMITED RESOURCE ALLOCATION

The scheduling procedure proposed for the solution of this problem is diagrammed in Figure 8-2; it represents a slight modification of the procedure outlined in Appendix 6-1 of the first edition of this text. This procedure will be described in conjunction with the network given in Figure 8-1. In this example, however, two resources, labeled A and B, will be considered, with requirements as given in columns 2 and 3 of Figure 8-3. Also given are the duration times (D), the initial early start (ES) and slack (S) times, and the late start times (LS). The procedure now steps through time scheduling of the ordered activities as soon as their predecessors are finished and sufficient resources are available.

To carry out this procedure two sets of activities are defined. First, the activities whose predecessors are all scheduled are called the Eligible

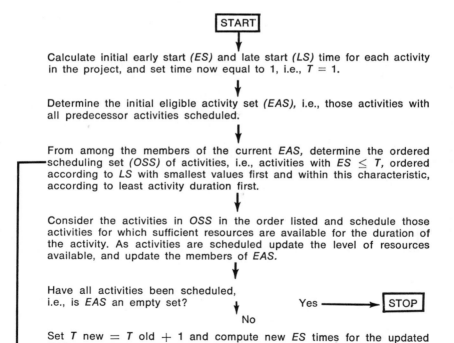

Figure 8-2   Basic multi-project multi-resource scheduling procedure.

Figure 8-3 Application of basic multi-project multi-resource scheduling procedure to network given in Figure 8-1.

Activity	Resource Req. A	B	D	ES	S	LS	1	2	3	4	5	6	7	8	9	10	11	12	13	14	15	16	17	18
0–1	3	—	2	1	5	6			x 3A	x 3A														
1–2	—	2	4	3	5	8								x 2B	x 2B	x 2B	x 2B							
0–3	6	—	2	1	0	1	x 6A	x 6A																
3–4	—	2	5	3	1	4			x 2B	x 2B	x 2B	x 2B	x 2B											
2–5	4	—	1	7	5	12														x 4A				
4–5	2	—	4	8	1	9								x 2A	x 2A	x 2A	x 2A							
0–6	3	—	1	1	6	7					x 3A													
3–7	4	4	8	3	0	3			x 4A 4B	x 4A 4B	x 4A 4B	x 4A 4B	x 4A 4B	x 4A 4B	x 4A 4B	x 4A 4B								
6–7	5	—	3	2	6	8											x 5A	x 5A	x 5A					
5–8	—	5	3	12	1	13															x 5B	x 5B	x 5B	
7–8	2	—	5	11	0	11														x 2A	x 2A	x 2A	x 2A	x 2A
Level of resource A unassigned							2	2	1	1	1	4	4	2	2	2	1	3	3	2	6	6	6	6
Level of resource A assigned						6 4 2																		
Level of resource B unassigned							6	6	2	2	0	0	0	0	0	0	4	6	6	6	1	1	1	
Level of resource B assigned						6 4 2																		

Activity Set (*EAS*). Now, since we are stepping through time, one unit at a time, we are only ready to consider activities with $ES \leqq T$. These activities are then ordered with least slack first and within this criterion with least duration first. This ordered list of activities is referred to as the Ordered Scheduling Set (*OSS*).

To simplify the bookkeeping in this scheduling procedure, the activities will be ordered according to their late start time (*LS*). The ordering obtained with this rule is identical to the ordering obtained by using slack (*S*). However, using *LS* values has the advantage that they do not change from time period to time period, whereas slack values continuously decrease for an activity that is ready to be scheduled but is not scheduled on any given day.

In the illustrative example, we start at $T = 1$ with the initial *EAS* being comprised of activities 01, 03, and 06. They are all members of *OSS*, because they are all eligible to start on the first "day" of the project. Ordering these activities according to late start times gives *OSS* as 03, 01, and 06, as indicated in Table 8-1. Since we are assuming in this example that a maximum of eight units of resource A and six units of resource B are available on each day of the project (note bottom of Figure 8-3), we can only schedule activity 03 on the first day. This activity is scheduled as shown in Figure 8-3 by the x's under days 1 and 2, and in the row for activity 03. The level of unassigned resources given at the bottom of Figure 8-3 is updated by marking the 8's down to 2's for resource A on days 1 and 2. The *EAS* in Table 8-1 is also updated by crossing through activity 03 and adding activities 34 and 37 whose only predecessor, 03, is now scheduled. Since the remaining resources will not permit any other activity to be scheduled on the first day, we are ready to progress to day 2 ($T = 2$), and update the corresponding *ES* values of the current *EAS* as shown in Table 8-1 under $T = 2$. The *ES* values for activities 34 and 37 are both 3, since their predecessor 03 is finished at the end of the second day. The ES values for the carryover activities 01 and 06 are now both 2, since they were not scheduled on day 1.

The complete schedule shown in Figure 8-3 can be obtained by continuing the above procedure as indicated in Table 8-1. The complete schedule given in Figure 8-3 indicates that the completion of the project is delayed three days past the early finish time of 15 computed in Figure 8-1 without regard to limitations on available resources. Another useful summary of this schedule is given by the resource loading diagrams at the bottom of Figure 8-3. They have been found to be quite useful in assessing the overall project resource requirements, and in making decisions about possible changes in resource availabilities.

Table 8-1. **Detailed Steps in Arriving at the Schedule Given in Figure 8-3**

$T=1$
EAS: 01 ~~03~~ 06 34 37
ES: 1 1 1
LS: 6 1 7

OSS: ~~03~~ 01 06
Schedule 03 to (1-2), remove 03
from EAS and add 34 and 37 to EAS

$T=2$
EAS: 01 06 34 37
ES: 2 2 3 3
LS: 6 7 4 3

OSS: 01 06
No activities can be scheduled
on $T=2$

$T=3$
EAS: ~~01~~ 06 ~~34~~ ~~37~~ 45 12
ES: 3 3 3 3
LS: 6 7 4 3

OSS: ~~37~~ ~~34~~ ~~01~~ 06
Schedule 37 to (3-10), remove 37
from EAS
Schedule 34 to (3-7), remove 34
from EAS and add 45 to EAS
Schedule 01 to (3-4), remove 01
from EAS and add 12 to EAS

$T=4$
EAS: 06 45 12
ES: 4 8 5
LS: 7 9 8

OSS: 06
No activities can be scheduled
on $T=4$

$T=5$
EAS: ~~06~~ 45 12 67
ES: 5 8 5
LS: 7 9 8

OSS: ~~06~~ 12
Schedule 06 to (5), remove 06
from EAS and add 67 to EAS

$T=6$
EAS: 45 12 67
ES: 8 6 6
LS: 9 8 8

OSS: 67* 12
No activities can be scheduled
on $T=6$

* 67 precedes 12 in OSS because
$D_{67} < D_{12}$

$T=7$
EAS: 45 12 67
ES: 8 7 7
LS: 9 8 8

OSS: 67 12
No activities can be scheduled on $T=7$

$T=8$
EAS: ~~45~~ ~~12~~ 67 25
ES: 8 8 8
LS: 9 8 8

OSS: 67 ~~12~~ ~~45~~
Schedule 12 to (8-11), remove 12
from EAS and add 25 to EAS
Schedule 45 to (8-11), remove 45
from EAS
(Note: Resource A is constraining
an activity with zero slack
and is thus causing schedule
to slip.)

$T=9$
EAS: 67 25
ES: 9 12
LS: 8 12

OSS: 67
No activities can be scheduled
on $T=9$

$T=10$
EAS: 67 25
ES: 10 12
LS: 8 12

OSS: 67
No activities can be scheduled
on $T=10$

$T=11$
EAS: ~~67~~ 25 78
ES: 11 12
LS: 8 12

OSS: ~~67~~
Schedule 67 to (11-13), remove 67
from EAS and add 78 to EAS

$T=12$
EAS: 25   78
ES: 12   14
LS: 12   11

OSS: 25
No activities can be scheduled
on $T=12$

$T=13$
EAS: 25   78
ES: 12   14
LS: 12   11

OSS: 25
No activities can be scheduled
on $T=13$

$T=14$
EAS: ~~25~~ ~~78~~ 58
ES: 14   14
LS: 12   11

OSS: ~~78~~ ~~25~~
Schedule 78 on (14-18), remove 78
from EAS
Schedule 25 on (14), remove 25
from EAS and add 58 to EAS

$T=15$
EAS: ~~58~~
ES: 15
LS: 13

OSS: ~~58~~
Schedule 58 on (15-17), remove 58
from EAS

$T=16$
EAS: Empty—STOP Scheduling
         Procedure

The above procedure illustrates a potentially powerful *approach* to a broad class of scheduling problems. First, it should be emphasized that while this example contained a single project network, the procedure is perfectly general and could be applied to any number of projects being conducted concurrently. With regard to multiprojects the only requirement of this procedure is that start dates and due dates are given for each project. This is required so that total activity slack (or the equivalent late start times) can be computed, which in turn determines the order in which the activities are considered for scheduling.* It should also be emphasized that the number of resources being considered is not limited by the procedure but only by the capacity of the computing system being used. Finally, if a large high-speed computer is available to carry out the steps in the scheduling procedure, the latter can be embellished considerably to simulate the many variations followed in the actual practice of scheduling project activities. These embellishments will be discussed below, after the problem of unlimited resource leveling is considered.

* Some researchers in this field, e.g., John Fondahl of Stanford University, use a slight modification of the procedure suggested in this text in the case of multiproject scheduling. That is, after each activity is scheduled the early finish time of the project is updated. If it exceeds the existing scheduled project completion time, the latter is replaced by the new early finish time. This new scheduled completion time is then used to update the late start times for unscheduled aciivities in the project in question. Unfortunately, there is no information available to the authors at this time to judge the merits of the addition of this step to the scheduling procedure proposed in this text.

UNLIMITED RESOURCE LEVELING

As mentioned above, the resource restrictions caused the project to run three days over the 15 day schedule which is possible if unlimited resources can be made available. Let us now consider this problem; what schedule will minimize the resource costs under the constraint that the project duration will not exceed 15 days? If the men are to be hired just for this project and discharged when the project is completed, then minimizing resource costs is equivalent to minimizing hiring and layoff costs, or putting it another way, leveling the resource requirements.

In the first edition of this text, a systematic procedure developed by Burgess[5] was presented for leveling resource requirements. This method utilized a measure of effectiveness given by the sum of the squares of the resource requirements for each "day" in the project schedule. It is easy to show that while the sum of the daily resource requirements over the project is constant for all complete schedules, the sum of the squares of the daily requirements decreases as the resource peaks are clipped to fill in valleys. Further, this sum reaches a minimum for a schedule which is level, or as nearly level as can be obtained for the project in question. The Burgess procedure is given in the following eight steps.

Burgess Leveling Procedure

STEP 1.

List the project activities in *order of precedence* by arranging the arrow head numbers in ascending order, and when two or more activities have the same head number, list them so that the arrow tail numbers are also in ascending order. (This assumes that the network events are numbered so that activity tail numbers are always less than the head numbers.) Add to this listing the duration, early start, and slack values for each activity as in Figure 8-4.

STEP 2.

Starting with the last activity (the one at the bottom of the diagram), schedule it to give the lowest total sum of squares of resource requirements for each time unit. If more than one schedule gives the same total sum of squares, then schedule the activity as late as possible to get as much slack as possible in all preceding activities.

STEP 3.

Holding the last activity fixed, repeat STEP 2 on the next to the last activity in the network, taking advantage of any slack that may have been made available to it by the rescheduling in STEP 2.

STEP 4.

Continue STEP 3 until the first activity in the list has been considered; this completes the first rescheduling cycle.

STEP 5.

Carry out additional rescheduling cycles by repeating STEPS 2 through 4 until no further reduction in the total sum of squares of resource requirements is possible, noting that only movement of an activity to the right (schedule later) is permissible under this scheme.

STEP 6.

If this resource(s) is particularly critical, repeat STEPS 1 through 5 on a different ordering of the activities which, of course, must still list the activities in order of precedence.

STEP 7.

Choose the best schedule of those obtained in STEPS 5 and 6.

STEP 8.

Make final adjustments to the schedule chosen in STEP 7, taking into account factors not considered in the basic scheduling procedure.

Considering again the example presented in Figure 8-3, suppose all activities are scheduled as early as possible. The results of this schedule are shown in Figure 8-4. The resource requirements for this schedule are quite irregular in time and require a maximum of 14 and 8 units of resources A and B, respectively. This schedule can obviously be improved. The application of the Burgess procedure will produce the results given in Figure 8-5 and summarized in Table 8-2 below. The details of obtaining Figure 8-5 from Figure 8-4 will be left as an exercise for the reader.

The schedule in Figure 8-5 could be improved by moving activity 34 one day to the left under a possible interpretation of STEP 8 of the above procedure. If this change in the schedule is made, the maximum level of resource B is reduced from 8 to 6, and the sum of squares from 375 to 367. This result will be compared with a second procedure developed by Wiest.[1]

Wiest Leveling Procedure

The inputs to this procedure are the same as for the Burgess procedure given above. The procedure consists of the following six steps:

STEP 1.

Schedule all jobs at early start, and plot manpower requirements in each shop for each day.

Figure 8-4 Resource loading with all activities scheduled at their early start times.

Activity	Resource Req. A	Resource Req. B	D	ES	S	LS	1	2	3	4	5	6	7	8	9	10	11	12	13	14	15
0-1	3	—	2	1	5	6	X 3A	X 3A													
1-2	—	2	4	3	5	8			X 2B	X 2B	X 2B	X 2B									
0-3	6	—	2	1	0	1	X 6A	X 6A													
3-4	—	2	5	3	1	4			X 2B	X 2B	X 2B	X 2B	X 2B								
2-5	4	—	1	7	5	12							X 4A								
4-5	2	—	4	8	1	9								X 2A	X 2A	X 2A	X 2A				
0-6	3	—	1	1	6	7	X 3A														
3-7	4	4	8	3	0	3			4A 4B	4A 4B	4A 4B	4A 4B	4A 4B	4A 4B	4A 4B	4A 4B					
6-7	5	—	3	2	6	8		X 5A	X 5A	X 5A											
5-8	—	5	3	12	1	13												X 5B	X 5B	X 5B	
7-8	2	—	5	11	0	11											X 2A	X 2A	X 2A	X 2A	X 2A
Level of resource A assigned							12	14	9	9	4	4	8	6	6	6	4	2	2	2	2
Level of resource B assigned									8	8	8	8	6	4	4	4		5	5	5	

Figure 8-5   Resource leveling using Burgess procedure.

Activity	Resource Req. A	Resource Req. B	D	ES	S	LS	1	2	3	4	5	6	7	8	9	10	11	12	13	14	15
0-1	3	—	2	1	5	6		x 3A	x 3A												
1-2	—	2	4	3	5	8								x 2B	x 2B	x 2B	x 2B				
0-3	6	2	2	1	0	1		x 6A													
3-4	—	5	5	3	1	4				x 2B	x 2B	x 2B	x 2B	x 2B							
2-5	4	1	1	7	5	12												x 4A			
4-5	2	4	4	8	1	9			x 3A	x 3A											
0-6	3	1	1	1	6	7			x 4A 4B	x 4A 4B	x 4A 4B	x 4A 4B	x 4A 4B	x 4A 4B							
3-7	4	8	8	3	0	3									x 4A 4B	x 4A 4B	x 2A	x 2A			
6-7	5	3	3	2	6	8					x 5A 5A	x 5A 5A	x 5A 5A								
5-8	—	5	3	12	1	13												x 2A 2A	x 5B 2A	x 5B 2A	x 5B 2A
7-8	2	—	5	11	0	11											x 2A	x 2A	x 2A	x 2A	x 2A
Level of resources A assigned	8	6	4	2			6	9	7	7	9	9	9	4	6	6	4	8	2	2	2
Level of resource B assigned	8	6	4	2			0	0	4	6	6	6	6	8	6	6	2	0	5	5	5

Table 8-2. Results of Resource Leveling on Figure 8-4

Resource Type	Early Start Time		Burgess Procedure		Wiest Procedure	
	Max. Res. Level	Sum of Squares	Max. Res. Level	Sum of Squares	Max. Res. Level	Sum of Squares
A	14	738	9	638	9	642
B	8	415	8 (6)*	375 (367)	6	367

*Figures in parentheses apply if activity 3-4 is scheduled in the interval (3-7).

STEP 2.

Calculate peak manpower requirements in each shop, and set "trigger levels" for all shops one unit below their respective peaks.

STEP 3.

Once again start scheduling jobs in technological order, calculating the manpower loading charts simultaneously. Stop when the trigger level of any shop (call it s) is exceeded.

STEP 4.

Examine the jobs that are active on the peak day in shop s. Compile a list of jobs which have sufficient slack to move them beyond the peak day without delaying the due date, and arrange them in descending order of their total slack. Pick one of these jobs (by a selection process* that favors the jobs highest on the list), and move it to the right on the Schedule Chart a random number of days between the minimum move necessary to push the job past the peak day and the maximum move allowed by its total slack.

STEP 5.

Continue with the scheduling of other jobs and plotting of the manpower loading chart. If additional peaks are generated, apply the procedure of (4). If all jobs are successfully scheduled, then lower the trigger levels of all shops one more unit and return to (3). If job shifting is not successful in removing peaks below the trigger levels, then restore the previous set of feasible trigger levels and attempt to reduce them shop by shop. As soon as no further reduction in trigger levels is possible, then print out the schedule.

STEP 6.

Repeat the above process (as many times as is computationally feasible). Because of the random elements in the program, it is likely that different schedules will result from each application of the program. Select as the final schedule the one having the lowest manpower costs (which are assumed to be proportional to the trigger levels—i.e., sufficient men are hired to meet peak loads and are paid whether idle or active on all days).

The application of this procedure, with a probability of selection in STEP 4 of $P = 1$, is shown in Figure 8-6. The results are summarized in Table 8-2 above and are very close, but not identical, to those

---

* The selection procedure contains random elements and operates as follows: with a probability of $P > 0$, select the first job in the list for the desired operation. If the first job is not selected, place it at the bottom of the list and select the second (now the top) job with the same probability P. Ultimately a job will be selected, as P is greater than zero. The probability of selecting any one job in repeated trials is a function of P and the number of jobs in the list, n. Thus the probability of selecting the ith job is $[P(1-P)^{i+1}/1 - (1-P)^n]$.

Figure 8-6 Resource leveling using Wiest procedure.

Activity	Resource Req. A	Resource Req. B	D	ES	S	LS	Time 1	2	3	4	5	6	7	8	9	10	11	12	13	14	15
0–1	3	—	2	1	5	6	3A	3A													
1–2	—	2	4	3	5	8			~~2B~~	~~2B~~	~~2B~~	~~2B~~		2B	2B	2B	2B				
0–3	6	—	2	1	0	1	6A	6A													
3–4	—	2	5	3	1	4			2B	2B	2B	2B	2B								
2–5	4	–	1	7	5	12							~~4A~~					4A			
4–5	2	—	4	8	1	9								2A	2A	2A	2A				
0–6	3	—	1	1	6	7			3A												
3–7	4	4	8	3	0	3			4A 4B	4A 4B	4A 4B	4A 4B	4A 4B	4A 4B	4A 4B	4A 4B					
6–7	5	—	3	2	6	8				5A	5A	5A									
5–8	—	5	3	12	1	13													5B	5B	5B
7–8	2	—	5	11	0	11											2A	2A	2A	2A	2A
Level of resource A assigned (trigger level = 9)							~~3~~ 9	~~3~~ 9	~~3~~ 7	~~4~~ 9	~~4~~ 9	~~4~~ 9	~~4~~ 8	~~2~~ 6	~~2~~ 6	~~2~~ 6	~~2~~ 4	~~4~~ 6	2	2	2
							x	x		x	x	x									
						8	x	x		x	x	x	x								
							x	x	x	x	x	x	x								
						6	x	x	x	x	x	x	x	x	x	x		x			
							x	x	x	x	x	x	x	x	x	x		x			
						4	x	x	x	x	x	x	x	x	x	x	x	x			
							x	x	x	x	x	x	x	x	x	x	x	x			
						2	x	x	x	x	x	x	x	x	x	x	x	x	x	x	x
							x	x	x	x	x	x	x	x	x	x	x	x	x	x	x
Level of resource B assigned (trigger level = 6)									~~2~~ ~~4~~ ~~8~~ 6	~~2~~ ~~4~~ ~~8~~ 6	~~2~~ ~~4~~ ~~8~~ 6	~~2~~ ~~4~~ ~~8~~ 6	~~2~~ 6	~~4~~ 6	~~4~~ 6	~~4~~ 6	2		5	5	5
						6			x	x	x	x	x	x	x	x					
									x	x	x	x	x	x	x	x			x	x	x
						4			x	x	x	x	x	x	x	x			x	x	x
									x	x	x	x	x	x	x	x			x	x	x
						2			x	x	x	x	x	x	x	x	x		x	x	x
									x	x	x	x	x	x	x	x	x		x	x	x

obtained by the Burgess procedure. It should also be noted that this procedure tends to schedule the activities as early as possible, which might be considered an advantage over the Burgess procedure, which tends to schedule the activities as late as possible. Finally, it should also be noted that the identical schedule given in Figure 8-6 could also be obtained by using the procedure given in Figure 8-2 with maximum resource levels equal to the trigger levels noted in Figure 8-6. This is an important point because it permits leveling, using a procedure that can be embellished to a considerable degree, as described in the next section.

EMBELLISHMENT OF THE BASIC SCHEDULING PROCEDURE IN FIGURE 8-2

While the procedure given in Figure 8-2 can be used as the basis of a general purpose scheduling procedure, a number of refinements are required before it can reach this status. Some of the possible modifications are described below, several of which are later incorporated in comprehensive general purpose scheduling procedures.

1. *Variable Crew Size.* Consider a construction job which would *normally* be carried out by three carpenters. If only two were available on a given day, we would probably start the job with the men we have; however, we may not be able to start with just one carpenter because of job requirements. Similarly, if we had more than three carpenters available, we might put four on the job, but not more than this. There are other types of activities where this type of variation may not be permitted. For example, the activity may require a skilled person or foreman, or heavy equipment such as cranes or drilling rigs that can only be scheduled at one level, i.e., on or off. In general, however, a job can be considered to have a normal manpower loading with a range from a minimum to a maximum level permitted if scheduling other than the normal number of men is desirable. In some cases, of course, the minimum and/or the maximum levels may be the same as the normal level.

2. *Splitting (Interrupting).* In most cases jobs can be split or interrupted, if necessary, to relinquish resources to satisfy a more critical demand, and it is overly restrictive not to permit this in some cases in a general purpose scheduling procedure. However, activities that depend on chemical or physical processes cannot be split, while activities that are critical or near critical, or involve heavy equipment, or those conducted by "subcontractors" should not be interrupted, once they have begun, for economic reasons.

A variation of splitting an activity is to reduce the resource loading short of zero, or short of interruption of the activity. If a currently scheduled (active) job utilizing a scarce resource at a level above its minimum allowable level has sufficient slack, it is often quite useful to permit borrowing men from this activity to permit the scheduling of a critical activity. Still another variation is to consider rescheduling the donor activity to a later start date.

3. *Assignment of Unused Resources.* Resource availabilities in a given category are usually given in terms of the number of men available on a "regular-time" basis, and in many cases an additional increment will be established that can be made available on an "overtime" basis, only if needed. After all possible activities have been scheduled on a given day, unassigned men from the regular-time group may remain in one or more resource categories. The usual practice is to allocate these unassigned resources, since they must be paid whether they are working or not, the allocation being made to active jobs that permit changes in resource levels during the conduct of the activity. To derive the most benefit from this practice, the assignment should be made to activities on the basis of least slack first.

4. *Project Priority Levels.* In the case of multi-project resource allocation, it often happens that project priority levels are given in addition to start dates and due dates. The optimal use of this information presents a problem. Scheduling the activities in the order of their priority levels is not desirable because this would hold up low priority activities unnecessarily, even though they might have negative slack. What we want to do is favor the high priority activities, but where these activities have sufficient slack, we want lower priority work, whose schedule is critical to move ahead and meet their due dates if possible. One way of accomplishing this is to transform high priority levels to reduced slack, and thereby favor the scheduling of these activities. A linear transformation of this type is given below for the case where priority levels go from a high priority, level 1, to a low priority, level 5.

$$\text{Modified Slack} = \text{Slack} - K\,(5 - \text{Priority Level})$$

In this equation K is a constant which must be chosen on the basis of experimentation. Using this equation, the highest priority project will have its slack reduced by $4K$ time units, whereas the adjustment is zero for the lowest priority level.

Some of the above embellishments of the basic scheduling procedure are incorporated in the procedure described in the next section to solve the long range resource planning problem.

LONG RANGE RESOURCE PLANNING

This problem has been described above as arising where management seeks to determine the combination of resource levels and project due dates that will minimize the total costs of resources, overhead, and losses which result when project due dates are not met. This is, indeed, a most difficult but important problem facing management of construction, maintenance, or engineering activities, to name a few of the potential applications.

Two procedures designed to solve this problem will be described in this section. They are both based on the ability of the high speed computer to economically simulate the effects of alternate resource availabilities on the ability of the firm to meet project due dates on future projects. The basis of these procedures are the basic types of scheduling procedures described above, with the addition of suitable embellishments to more nearly simulate actual company scheduling practices.

Wiest's SPAR-1 Model

One of the more recently developed comprehensive scheduling procedures, due to Wiest,[1,12] is called SPAR-1 (Scheduling Program for Allocating Resources). The flow diagram for this procedure is given in Figure 8-7. It is similar to the one presented in Figure 8-2 in that it is based on the assumption that each project is described by an arrow diagram with an assigned start date and due date, that the project activities are ordered for scheduling according to least slack first, and that the actual scheduling takes place one time unit at a time from the first to the last. However, SPAR-1 embodies a number of useful embellishments. First, resource requirements for each activity or job include a maximum and a minimum loading, in addition to a normal loading, if different levels are permissible. The total man-days required for the job is assumed to be constant for all resource loadings. In cases where it is needed, the job data will also include a restriction requiring a resource loading to be maintained constant at its initial level, during the life of a job.

The SPAR-1 model begins by computing ES and S for each job. Then jobs are scheduled, day by day, starting with day $d = 1$, by selecting jobs with $ES = d$ and ordered according to their slack. Jobs are scheduled as available resources permit. If a job cannot be scheduled on day d, then it is updated so that $ES = d + 1$ and its slack is reduced by

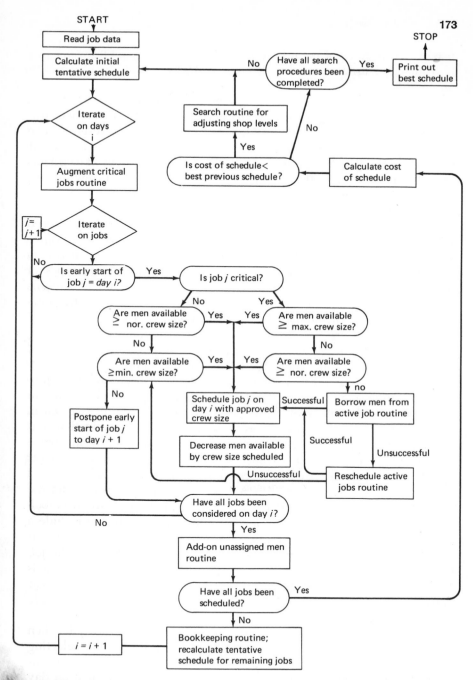

Figure 8-7 Flow diagram for SPAR-1 scheduling program for allocating resources.

one day. This much of the procedure is the same as Figure 8-2. However, also included in the model is a random device which enables one to obtain different schedules for the same project inputs. With a probability of $P > 0$, the first job in the list is selected for scheduling. If the first job is not selected, it is placed at the bottom of the list and the second job, now at the top of the list, is selected with the same probability P. Eventually all of the jobs in the list will be selected, but the order of scheduling the jobs is varied randomly, and hence the schedule itself is varied. The speed of the computer is then exploited to generate a number of complete schedules from which the best, according to a defined criterion, is selected.

Referring to Figure 8-7, we see that the first consideration, as scheduling is started on a new day, d, is to augment critical jobs that were started before day d and will be active beyond day d. These jobs must not already be loaded to their maximum resource level, and they must permit variation in their resource loading from day-to-day. After this routine is completed, new jobs are considered for scheduling as described above, including the random selection device. If the job being considered for scheduling is not critical, attempts are made to schedule it with a normal crew size. If this is not possible, the minimum crew size is considered before the job is postponed until day $d + 1$. If the job being considered is critical, attempts are made to schedule it with maximum crew size, or if this is not possible, a normal crew size is considered. If neither of these attempts are successful, two special subroutines are tried. First, efforts are made to borrow the needed resource from currently active jobs that will not become critical if their rate of progress is reduced. If necessary, a second subroutine is called which reschedules active jobs to a later start date. If both of these attempts fail, a minimum crew size schedule is considered before the start of the critical job is postponed. After all of the jobs that can start on day d have been considered for scheduling, any remaining unassigned resources are added on to currently active jobs.

After all jobs in the project(s) have been scheduled, the procedure enters the stage where long range planning problems are considered. First, the cost of the schedule just completed is computed, considering daily overhead expenses, due date penalties and possibly bonuses, and resource costs based on the assumed available levels. Depending on the cost of the schedule compared with previous schedules, the procedure is either terminated, or a new set of available resources is computed and tested.

The quantitative basis for carrying out the various steps in the SPAR-1 model is given in reference 1. It has been programmed for second generation computers which are capable of handling moderately large size problems. On third generation computers, one should be able to handle quite large problems, involving several thousand or more activities and any reasonable number of resources. The results obtained from the SPAR-1 model on several test projects that were suitably restricted so optimal schedules could be determined, indicated that it gives very nearly optimal results. For actual projects where optimal schedules could not be ascertained, the SPAR-1 schedules were considered to be definite improvements over those obtained by existing methods.

RAMPS—RESOURCE ALLOCATION AND MULTIPROJECT SCHEDULING

One of the earliest comprehensive resource scheduling procedures addressed to the long range resource planning problem is called RAMPS.[8,14] Although its development was entirely independent of the SPAR-1 model, RAMPS is also based on an arrow diagram representation of each project, together with three resource loading and corresponding time estimates for each activity. Unlike SPAR-1, which assumes constant resource efficiency at all levels of resource loading, RAMPS permits variation in the total man-hours for the three levels. Also included in RAMPS is the cost of interrupting (splitting) a job once it has begun.

At the project level, the input information includes the start date, desired completion date, and dollar-penalty rate for delay of completion, or as an alternative, a project priority rating. With regards to resource availabilities, the input information must give, for each time period, the normal cost and available number of units of each type of resource, and the extra number of units and their cost, which may be made available through overtime and subcontracting. Finally, scheduling objectives must be stated in terms of relative importance (weights) of minimizing idle resources, meeting project completion dates, avoidance of activity interruption (splitting), maximizing the number of activities scheduled concurrently, etc. After the basic scheduling computations are made using all normal times, this program progresses through the network, the activities being time scheduled as long as resources are available. If the available resources are not sufficient, the various feasible combinations of allocations are evaluated on the basis of cost, either direct or implied, and the minimum cost combination is chosen. The rules under which costs are associated with each combination reflect the relative weights given to the various scheduling objectives.

There are two main outputs of this program; one gives the individual activity schedules, costs, and resources, summarized by projects, and the other gives the resources used by type and time period summarized over all of the projects. A study of these outputs usually suggests certain changes in the inputs that will bring the former more in line with desired objectives, whatever they may be. For example, certain projects may not be completed on time. Assuming the desired completion date is feasible, the computer output will indicate the resource bottleneck causing the project completion delay. Similarly, if resource idleness is excessive, the effects of a particular reduction in resource availability can be determined by rerunning the program at reduced resource levels. Evidently, a few such computer runs will, in most cases, lead to an acceptable master schedule, which is updated periodically to accommodate changes in current plans, cancellation and completion of current projects, and the introduction of new projects.

The above brief descriptions of SPAR-1 and RAMPS indicate that they contain a considerable number of similarities in data handling ability and in their general approach to scheduling. There are, however, several basic differences worth noting. First, on each day that is scheduled, RAMPS schedules resource by resource starting with the one in most critical demand. Each resource has a "criticality index," which is based on the total man-days required for a resource summed over all jobs and all days, and the man-days of that resource available. The program generates, for each resource on each day, all nontrivial assignment patterns which are then evaluated considering slack, number of jobs scheduled, number of jobs split, idle resources, and criticality (in terms of resource needs) of immediate successor jobs. The pattern with the highest score is selected, and the program proceeds to the next resource, or to the next day. Once a job is scheduled, it remains so. Thus, RAMPS does considerable computing in arriving at a days schedule, thereby minimizing costs each day. SPAR-1, on the other hand, does less computing each day, but attempts to optimize over the entire schedule by using add-on, borrow, reschedule and search routines. One can develop network problems which can best be handled by one or the other of these two procedures; each has advantages not shared by the other.

SUMMARY

In this chapter we have treated the problem of resolving resource constraint problems, using the arrow diagram representation of a project plan as the underlying basis of each of the suggested procedures. It was shown that simple resource constraints could be resolved by trial and error scheduling of activities on slack paths. This type of exercise only requires the basic early and late start and finish times, and total slack times, which can be conveniently displayed on the project network as developed in Chapter 4.

More involved questions, dealing with multiprojects and multi-resources, require formal procedures that can be programmed for computer processing. Even with a computer, however, the complex combinatorial nature of the resource allocation problem usually precludes the objective of obtaining an optimal solution. However, we will undoubtedly see continuing improvements to current procedures, such as the one developed by Davis,[3] seeking optimal solutions under restricted assumptions.

While algorithms for optimal solutions to the general scheduling problem offer little promise with the current state of computational resources, the difficulty of the problem and its implications is too formidable to be adequately handled entirely "by hand." Heuristic scheduling rules, programmed to give "good" schedules, have been the basis for practical working systems developed to date. This approach, however, makes very limited use of the "imagination" available to the planner himself and often must make necessarily naive assumptions as to his goals and alternatives. This problem must be considered seriously in the future, in view of the ever-increasing ability to communicate with the computer itself in a direct and expeditious way. The widespread availability of remote or local access to a computer leads one to speculate about the not too distant future when the planner himself can suggest alternatives for exploration in real time, and allow the machine to rapidly compute all the consequential implications. This approach would be a true exploitation of the heuristic approach in which the machine would be a very valuable tool, serving to amplify rather than supplant the technicians' imagination.

REFERENCES

1. Wiest, Jerome D., "Computer Models for the Scheduling of Large Projects." Ph.D. Dissertation, Carnegie Institute of Technology, September, 1964.
2. Davis, Edward W., "Resource Allocation in Project Network Models—A Survey," *Journal of Industrial Engineering*, Vol. XVII, No. 4, April, 1966, pp. 177-188.
3. Davis, Edward W., "An Exact Algorithm for the Multiple Constrained-Resource Project Scheduling Problem." Ph.D. Dissertation, Yale University, May, 1968.
4. Clark, C. E., "The Optimum Allocation of Resources Among the Activities

of a Network," *Journal of Industrial Engineering*, January–February, 1961.

5. Burgess, A. R., and J. B. Killebrew, "Variation in Activity Level on a Cyclical Arrow Diagram," *Journal of Industrial Engineering*, Vol. 13, No. 2, March–April 1962.

6. McGee, A. A. and Markarian, M. D., "Optimum Allocation of Research/Engineering Manpower Within a Multi-Project Organizational Structure," *IEEE Transactions on Engineering Management*, September, 1962.

7. Kelley, J. E., "Scheduling Activities to Satisfy Constraints on Resources," *Industrial Scheduling*, Muth and Thompson, Editors, 1963.

8. Moshman, J., Johnson, J., and Larsen, M., "RAMPS, A Technique for Resource Allocation and Multiproject Scheduling," *Proceedings, 1963 Spring Joint Computer Conference.*

9. Mize, J. H., "A Heuristic Scheduling Modeo for Multiproject Organizations" Unpublished Doctoral Dissertation, Purdue University, August, 1964.

10. Conway, R. W., "An Experimental Investigation of Priority Assignment in a Job Shop," Memorandum RM-3789-PR, The RAND Corporation, February, 1964.

11. Brooks, G. H. Unpublished paper, Purdue University. See Appendix 6-1 of 1st Edition, *Project Management with CPM and PERT*, Reinhold Publishing Co., New York, 1964.

12. Wiest, Jerome D., "A Heuristic Model for Scheduling Large Project with Limited Resources," *Management Science*, Vol. 13, No. 6, February 1967, pp. B359-B377.

13. Fendley, Larry G., "Toward the Development of a Complete Multiproject Scheduling System," *Journal of Industrial Engineering*, Vol. 19, No. 10, October 1968, pp. 505-515.

14. Lambourn, S., "Resource Allocation and Multiproject Scheduling (RAMPS) —A New Tool in Planning and Control," *The Computer Journal*, Vol. 5, No. 4, January, 1963, pp. 300-304.

EXERCISES

1. Suppose you are planning to use the same set of concrete forms on certain of the activities making up the project networked in Figure 8-1.
   a. If the forms are needed on activities 0-3, 0-1, 2-5, and 7-8, can they be scheduled so that only one set of forms are required, and so that the project is not delayed? If so, give all possible schedules.
   b. Suppose the forms are needed on activities 0-3, 4-5, and 7-8. Find the schedule which causes the least amount of delay in completing the project.
   c. If the requirements of one set of forms as described in (b) above had been considered in the planning phase of this project, how would it have been incorporated in the project network?

2. Repeat the illustrative Limited Resource Allocation example given in Figure

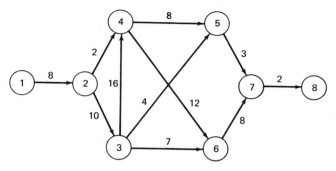

Figure 8-8

8-3, using the following revised list of resource requirements. Show that this set of resource requirements requires a 21 day project duration.

Activity	A	Resource Requirement B	C*
0-1	—	3	1
1-2	—	2	1
0-3	3	—	1
3-4	—	2	1
2-5	4	—	1
4-5	2	—	1
0-6	2	—	1
3-7	4	4	1
6-7	5	—	1
5-8	—	5	1
7-8	2	—	1
Maximum Level Available	6	6	2

*Resource C represents foremen, and only two are available for assignment to this project.

3. Verify the project schedule shown in Figure 8-5 using the Burgess leveling procedure.

4.*Consider the network shown in Figure 8-8, which represents one cycle of work which is to be repeated many times in a project. Since the longest activity in this cycle has a duration of 16 days, assume that the cycle is to be repeated at 16-day intervals. If each activity in the network requires one crew of the type under consideration, find the optimal cycle schedule of activities, which in this case requires 5 crews on each of the 16 days which make up a cycle.

5. Verify the project schedule shown in Figure 8-6 using the Wiest leveling procedure, omitting the random selection feature of this procedure.

* This exercise is taken from the paper by Burgess.[5]

6. Rework the example presented in the text in Figure 8-3 using the Wiest procedure outlined in Figure 8-7. To carry this out, assume that the crew requirements stated in Figure 8-7 are normal loadings, and the maximum and minimum loadings are one crew more or one crew less, respectively.
   a. Write out in precise, programmable terms, the way you would carry out the various scheduling features of the Wiest scheduling procedure.
   b. Apply the above procedure to the modified example in Figure 8-3, assuming resource availabilities of eight and six crews for resource *A* and *B* respectively.

9

TIME-COST TRADE-OFF PROCEDURES

The results of the planning and scheduling stages of the critical path method provide a network plan for the activities making up the project and a set of earliest and latest start and finish times for each activity. In particular, the earliest occurrence time for the network terminal event is the estimated "normal" project duration time, based on "normal" activity time estimates. This state of the overall project planning and control procedure is depicted in box (3) of Figure 1-4 (Chapter 1). The purpose of this chapter is to consider the question raised in the next step, i.e., whether the current plan satisfies time constraints placed on the project.

Time constraints arise in a number of ways. First, the "customer" might contractually require a scheduled completion time for the project. Then, the original time constraint might change after a project has started, requiring new project planning. These changes arise because of changes in the customer's plans; or, when delays occur in the early stages of a project, the new expected completion time of the project may be too late. The most interesting time constraint application, and the one which was the basis for the development of the CPM time-cost trade-off procedure by Kelley and Walker,[1,2] arises when we ask for the project schedule that just balances the value of time saved against the incremental cost of saving it. This situation occurs frequently, for example, in the major overhaul of large systems, such as chemical plants, paper machines, aircraft, etc. Here the value of time saved is very high, and furthermore, it is known quite accurately. In this application, the crux of the problem amounts to developing a procedure to find the minimum cost of saving time. This assumes, of course, that some jobs can be done more quickly if more resources are allocated to them. The resources may be men, machinery, and/or materials. We will assume that these resources can be measured and estimated, reduced to monetary units, and summarized as a cost per unit time.

Thus, the main purpose of this chapter can be stated as the development of a procedure to determine schedules *to reduce the project duration time with a minimum increase in the project direct costs, by buying time along the critical path(s) where it can be obtained at least cost.*

This procedure can be applied informally in a very simple manner. For example, consider the network frequently used in this text, shown in Figure 9-1. The critical path is composed of activities 0-3, 3-7 and 7-8, with normal duration times of 2, 8, and 5 weeks, respectively, giving a project duration of 15 weeks. Also given on the network is the cost of buying time on the critical path. It is easy to see in this example that the cheapest way to compress this project is to add "resources" to activity 3-7 at a cost of $250, and reduce its duration from 8 to 7 weeks. The net result is shown by the updated computations which indicates a compressed project duration of 14 weeks. At this point, it is interesting to note that there are now two critical paths and further reductions must consider both of them.

Systematic methods of carrying out the above procedure will be taken up in this chapter. First, the original CPM approach of Kelley and Walker,[1,2] based on simple linear time-cost trade-off curves for each activity, will be presented. A structural model, due to Prager,[3] will be used to give insight to this procedure. Then, a heuristic hand computational procedure will be presented which can handle very general time-

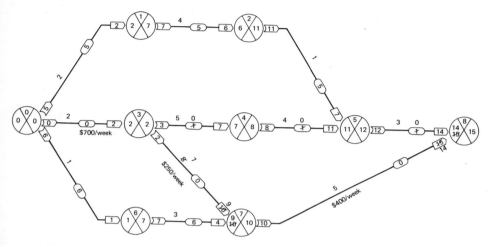

Figure 9-1 Example of elementary time-cost trade-off procedure showing compression of project from 15 to 14 days.

cost trade-off relationships for the project activities. Finally, an extensive treatment of the use of linear programming to give exact solutions to this problem is presented in Appendix 9-1.

It is assumed in all of the procedures to be developed in this chapter that unlimited resources are available. If this is not the case, or if personnel must be paid up to some maximum resource requirement level, whether they are needed or not on a particular day, then the procedures described in Chapter 8 under "Long Range Resource Planning" may be more appropriate than those described here. Although this assumption of unlimited resources is not often completely satisfied, there are a number of situations where it is satisfied to the extent required here. For example, the project in question may be a high priority project, which will draw personnel from a large number of low priority or deferrable work activities, so that there are effectively unlimited resources.

THE CRITICAL PATH METHOD (CPM) OF TIME-COST TRADE-OFFS

The development of the basic CPM time-cost trade-off procedure is based on a number of special terms which are defined below and are further shown in Figure 9-2.

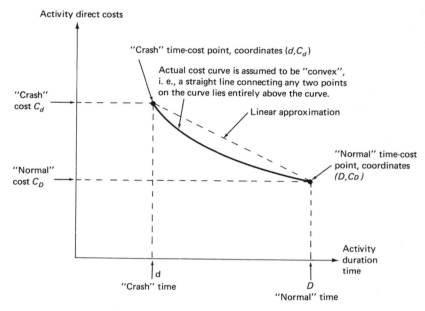

Figure 9-2   Activity time-cost trade-off input for the CPM procedure.

Definition:

Activity direct costs include the costs of the material, equipment, and direct labor required to perform the activity in question. If the activity is being performed in its entirety by a subcontractor, then the activity direct cost is equal to the price of the subcontract.

Definition:

Project indirect costs may include, in addition to supervision and other customary overhead costs, the interest charges on the cumulative project investment, penalty costs for completing the project after a specified date, and bonuses for early project completion.

Definition:

Normal activity time-cost point. The normal activity cost is equal to the absolute minimum of direct costs required to perform the activity, and the corresponding activity duration is called the normal time. (It is this normal time that is used in the basic critical path planning and scheduling, and the normal cost is the one usually supplied if the activity is being subcontracted.) The normal time is actually the shortest time required to perform the activity under the minimum direct cost constraint, i.e., this rules out the use of overtime labor or special time saving (but more costly) materials or equipment.

Definition:

Crash activity time-cost point. The crash time is the fully expedited or minimum activity duration time that is possible, and the crash cost is assumed to be the minimum direct cost required to achieve the crash performance time.

The normal and crash time-cost points are denoted by the coordinates (D, C_D) and (d, C_d), respectively, in Figure 9-2. For the present, it will be assumed that the resources are infinitely divisible, so that all times

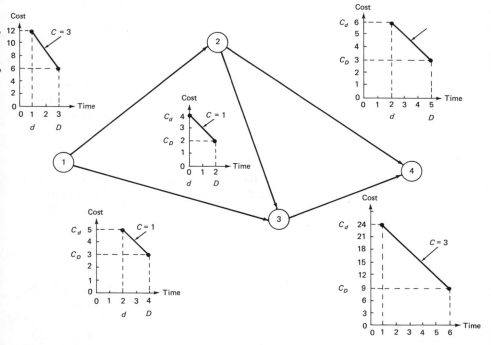

Figure 9-3 Illustrative network giving activity time-cost trade-off data for the CPM procedure.

between d and D are feasible, and the time-cost relationship is given by the solid line. It will also be assumed that this curve is convex, as shown in Figure 9-2, and can be adequately approximated by the dashed straight line. These assumptions will be relaxed in subsequent procedures.

To explain the basic time-cost trade-off principle, a rather ingenious analogy devised by Prager[3] will be used on the simple network given in Figure 9-3. The activity time-cost trade-off curves are given beside each activity in this figure. For example, activity 1-3 has a normal performance time of $D_{13} = 4$ days, with a corresponding normal direct cost of $C_{D13} = \$3$, and a crash or minimum time of $d_{13} = 2$ days, with a corresponding direct cost of $C_{d13} = \$5$. It is assumed that all times between 2 and 4 days are feasible; however, since integer times will be used throughout the network, only integer times will be assigned to the activities in any of the problem solutions. It is also assumed that the cost relationship can be adequately approximated by a straight line with a cost slope of $C_{13} = \$1/\text{day}$. Each of the other four activities are described by a similar time-cost trade-off curve.

The mechanical analogy to be used here is explained in Figure 9-4 using activity 1-3 for illustrative purposes. Each activity (i, j) in the network is represented by a structural member that consists of a rigid sleeve of length d_{ij} (crash time), containing a compressible rod whose natural length is D_{ij} (normal time), with a piston at its protruding end. When the member is subjected to a gradually increasing compressive

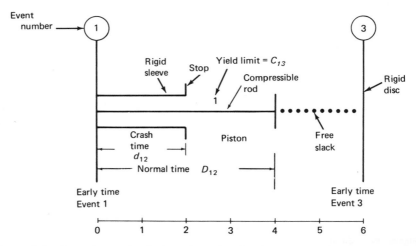

Figure 9-4 Prager's mechanical analogy for the time-cost trade-off problem.

force, f_{ij}, the rod remains rigid until this force reaches intensity C_{ij} (cost slope), which will be referred to as the "yield limit" of the member i–j. A force of this intensity can freely compress the rod, but the piston and a stop at the end of the sleeve make it impossible for the rod to be compressed by more than the difference $D_{ij} - d_{ij}$. When the rod has been compressed by this amount, any further increase of the compressive force applied to the member will be carried by the sleeve.

In the structural model, the members representing the activities of

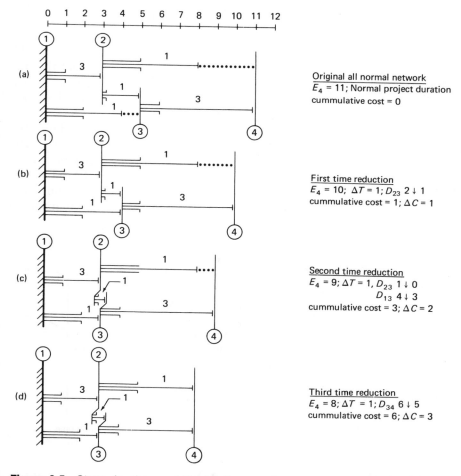

Figure 9-5 Steps in time-cost trade-off procedure for the network given in Figure 9-2.

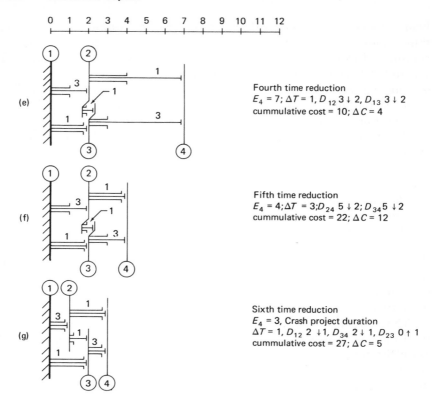

(e)

Fourth time reduction
$E_4 = 7; \Delta T = 1, D_{12} 3 \downarrow 2, D_{13} 3 \downarrow 2$
cummulative cost = 10; $\Delta C = 4$

(f)

Fifth time reduction
$E_4 = 4; \Delta T = 3; D_{24} 5 \downarrow 2; D_{34} 5 \downarrow 2$
cummulative cost = 22; $\Delta C = 12$

(g)

Sixth time reduction
$E_4 = 3$, Crash project duration
$\Delta T = 1, D_{12} 2 \downarrow 1, D_{34} 2 \downarrow 1, D_{23} 0 \uparrow 1$
cummulative cost = 27; $\Delta C = 5$

Figure 9-5   Continued

the project are arranged between thin rigid discs, which represent the network events. Figure 9-5 shows the model for the project network shown in Figure 9-3. Figure 9-5a shows a configuration that corresponds to the normal project duration of 11 days. This can be noted by the location of the thin disc for event 4, which is at 11 on the time scale. The reader should be able to identify each of the five project activities in Figure 9-5a. Note also the free slack which occurs at merge events 3 and 4. The two paths leading to event 3 are 1-2-3 and 1-3, of length 5 and 4, respectively. Hence, activity 1-3 has one day of free slack, shown by the dotted line from the piston to the disc representing event 3. In a similar manner, we find 3 days of free slack for activity 2-4 at the merge event 4. The reader should also be able to see the critical path composed of the rods representing activities 1-2, 2-3, and 3-4.

To "compress" the project duration time, disc 1 is held fixed, and a gradually increasing compressive load L is applied to disc 4, all discs being guided in such a manner that they can freely perform horizontal translations. When the load reaches 1 unit on disc 4, the force will be transmitted through the rod 3-4 to disc 3, which in turn, will compress the rod 2-3, which has a cost slope of \$1/day. This compression will stop after a travel of 1 day, at which time discs 3 and 4 will have moved 1 day to the left, as shown in Figure 9-5b. Thus, we have compressed the duration of the project from 11 to 10 days, at a minimum increase in the direct costs of \$1/day. We also note that there are now two critical paths, i.e., 1-2-3-4 and 1-3-4.

At this point, compression in the project duration must be along activity 3-4, which is common to both critical paths, or along activity 1-3 on one path and either 1-2 and 2-3 on the other path. It turns out that movement of disc 4 will take place when the load is increased to 2 units, at which time rods 2-3 and 1-3 will compress simultaneously. Movement under this load will cease when discs 3 and 4 have again moved one day to the left as shown in Figure 9-5c. Now we have compressed the duration of the project from 10 to 9 days, at an incremental cost increase of \$2/day, for a total cost increase of \$3 for the two days. There continue to be two critical paths, i.e., 1-2-3-4 and 1-3-4.

The compression of the project duration continues in this manner until the final compression which takes place in going from 4 to 3 days. With the project duration at 4 days, as shown in Figure 9-5f, we have three critical paths, i.e., 1-2-4, 1-2-3-4, and 1-3-4. This presents an interesting situation. To compress the project as shown in Figure 9-5f, it is clear that both activities 1-2 and 3-4 must be compressed, each requiring a force of 3 units. However, activity 2-3 has already been compressed at a load of 1 unit, which is now acting to compress activity 1-2. As a result, a total load of 5 units will compress the project from 4 to 3 days at a net cost of \$5/day. This cost is made up of \$3/day on activity 1-2, \$3/day on activity 3-4, and a negative \$1/day on activity 2-3, where time has literally been sold back, giving a net cost of \$5/day. It should be noted that of the activities 1-2 and 3-4 that were compressed here, one lies along path 1-2-4, one along path 1-3-4, and two along path 1-2-3-4. The latter fact gives rise to the selling back of time previously bought on activity 2-3.

A summary of the above compression steps is given in Table 9-1. Here, it has been assumed for illustrative purposes that the indirect cost rate is constant at \$3/day. These same results are shown in Figure 9-6, which shows that the optimum project duration, from a minimum total cost standpoint, is 8 or 9 days. The corresponding schedule of the project activities can be seen in the corresponding Figures 9-4c or 9-4d. This result of two different minimum cost schedules occurs because the

Table 9-1. Summary of Steps in Time-Cost Trade-Off Shown in Figure 9-5

Time Reduction	Project Duration	Total Direct Costs	Total* Indirect Costs	Total Costs	Activity Changes
All-Normal Schedule	11	23	33	56	—
First	10	24	30	54	$y_{23} \rightarrow$ 2 to 1
Second	9	26	27	53	$y_{23} \rightarrow$ 1 to 0; $y_{13} \downarrow$ 4 to 3
Third	8	29	24	53	$y_{34} \rightarrow$ 6 to 5
Fourth	7	33	21	54	$y_{12} \rightarrow$ 3 to 2; $y_{13} \rightarrow$ 3 to 2
Fifth	4	45	12	57	$y_{24} \rightarrow$ 5 to 2; $y_{34} \rightarrow$ 5 to 2
Sixth	3	50	9	59	$y_{13} \rightarrow$ 2 to 1; $y_{34} \rightarrow$ 2 to 1; $y_{23} \leftarrow$ 0 to 1

*Indirect costs are assumed to be $3/day.

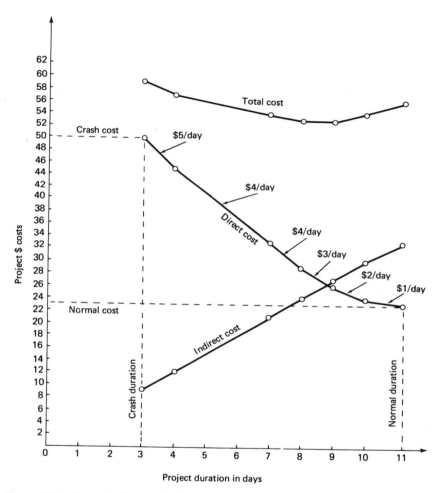

Figure 9-6 Direct, indirect and total project costs as a function of project duration for the network in Figure 9-2.

third time reduction of $3/day is exactly equal to the value of time, i.e., the assumed indirect cost of $3/day.

In this very simple example, the compression of the structure by a gradually increasing load could be followed intuitively. For more realistic examples, however, a general algorithm is needed that clearly indicates which members are deforming and which remain rigid during each phase of the compression process. The network flow algorithm, developed for this purpose by Fulkerson,[4] is given in Appendix 9-1; its application is illustrated on the same network used above. This algorithm is a rigorous and highly efficient computational procedure which has been programmed for computer processing to give the *optimum (minimum)* direct cost schedules for total project duration times ranging from the normal time to the shortest possible or crash time, as shown in Figure 9-6. Computer programs for this purpose are described in Chapter 5.

ACTIVITY TIME-COST TRADE-OFF INPUTS FOR THE CPM PROCEDURE

The basic activity inputs to the CPM procedure have been described above, and illustrated in Figure 9-2, where it is assumed that the time-cost trade-off points lie on a *continuous linear or piece-wise linear* decreasing curve. A piece-wise linear curve is illustrated in Figure 9-7. It is further assumed that the activities are independent, in the sense that buying time on one activity does not affect in any way the availability, cost, or need to buy time on some other activity. This assumption would, for example, be violated if a special resource could be obtained to speed up simultaneously two separate activities in the network, since this would mean that buying time on one activity would automatically include the other.

If the *actual* time-cost relationship departs significantly from the assumed straight line, but is "convex," then it may be necessary to fit the actual cost curve with a series of straight lines, as shown in Figure 9-7. A convex curve is one for which a straight line connecting any two points on the curve lies entirely above the curve; similarly, if the straight line lies entirely below the curve, it is said to be "concave." These definitions are illustrated in Figure 9-8; the need for the convexity assumption will be described below.

In Figure 9-7, the actual cost curve, which is convex, is approximated by a piece-wise linear curve, each piece being treated as a separate activity or pseudo-activity. In the project network, the actual activity, A, is replaced by the pseudo-activities, A_1, A_2, and A_3 drawn in series as shown in Figure 9-7. In this illustration, the approximation is with three pieces; however, the procedure can be extended in an obvious way to any number of pieces. The coordinates of the normal and crash time-cost points for each pseudo-activity are given in the table at the bottom of the figure, where it can be noted that the sum of the three pseudo-activities, A_1, A_2, and A_3, gives the whole activity, A, and the sum of the coordinates of the normal and crash points for the pseudo-activities gives

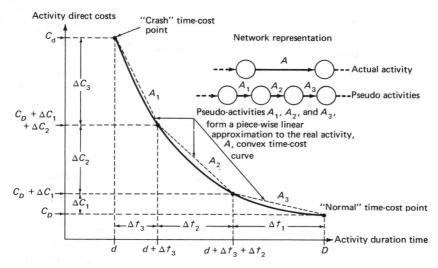

|Cost slope| = ("crash" cost - "normal" cost) / ("normal" time - "crash" time)

Pseudo-activities	"Normal"		"Crash"		Time-cost slope
	Time	Cost,$	Time	Cost,$	
A_1	$d + \Delta t_3$	0	d	ΔC_3	$(\Delta C_3 / \Delta t_3)$
A_2	Δt_2	0	0	ΔC_2	$(\Delta C_2 / \Delta t_2)$
A_3	Δt_1	C_D	0	$(C_D + \Delta C_1)$	$(\Delta C_1 / \Delta t_1)$
Total: A	$D =$	C_D	d	$C_d =$	$(C_d - C_D)/(D - d)$

$$D = d + \Delta t_3 + \Delta t_2 + \Delta t_1 \qquad\qquad C_d = \$ (C_D + \Delta C_1 + \Delta C_2 + \Delta C_3)$$

Figure 9-7 Piece-Wise linear approximation to convex time-cost curves using pseudo-activities.

the coordinates of the same points for the whole activity, i.e., (D,C_D) and (d,C_d). The reason for the convexity requirement can be explained heuristically in terms of Figure 9-7. If the activity is currently scheduled at its normal time, D, then physically, pseudo-activity A_3 must be aug-

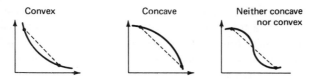

Figure 9-8 Illustration of convexity and concavity.

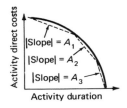

Figure 9-9 Example of a concave activity time-cost trade-off curve.

mented first, then A_2, and finally A_1. Since the CPM computational procedure effectively searches the critical activities to find the one that can be augmented the cheapest, it will naturally choose the pseudo-activities in the proper order, i.e., A_3, A_2, and finally A_1, since the cost slopes increase as one goes from A_3 to A_2 to A_1, for any *convex* curve. However, if the time-cost curve was not convex, then the cost slopes may be lowest for A_1, and highest for A_3, as shown in Figure 9-9. In this case, the CPM computational procedure would augment the activities in a sequence that would not be physically meaningful, i.e., in the order A_1, A_2, and finally A_3.

The CPM Computational Procedure

The CPM computational procedure chooses the duration times for each activity so as to minimize the total project direct costs and at the same time satisfy the constraints on the total project completion time and on the individual activities, the latter being dictated by both the logic of the project network and the performance time intervals (d,D) established for each activity.

Just as each activity performance time is constrained to some interval (d,D), the total project completion time is constrained in the interval denoted by (T_d, T_D). If one schedules each activity to be performed at its normal time, one will have the lowest possible direct cost schedule which will, however, require the longest scheduled time, T_D, to complete—point A in Figure 9-10. Similarly, a minimum project duration time, T_d, can be achieved by utilizing the most costly "crash" activity times. If the latter are employed only where necessary, i.e., on the final critical path(s), the total project time-cost point is indicated by B. The use of crash activity times *across the board* increases the total project direct costs from point B to point C; however, the extra resources are wasted, since no further reduction in the project duration is possible. One could also consider the most unwise procedure of crashing all activ-

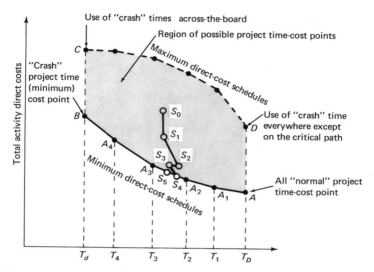

Figure 9-10   Region of all possible project time-cost points.

ities except those on the original critical path(s) which would be scheduled at their normal times. This would lead to point *D*, which corresponds to a project duration of T_D, but a much higher total project direct cost than that resulting from the use of all normal times. Finally, the curve *DC* is obtained by starting with the schedule corresponding to point *D*, and progressively augmenting combinations of activities on the critical paths with the highest cost slopes.

The cross-hatched area, *ABCD*, in Figure 9-10 has been developed merely to show the entire region of possible project time-cost points that could be scheduled. The CPM computational procedure, however, is ordinarily only used to develop the minimum project direct cost curve *A-B*, which is a series of straight lines, and the detailed project schedules corresponding to the points *A*, A_1 A_2, A_3, A_4, and *B*. The cross-hatched area shows how far a schedule can depart from the minimum cost schedule curve for a specific project duration. It shows, in particular, at point *C* how uneconomical can be the practice of speeding up a project by using crash times *across the board*, instead of along the critical paths only.

Specifically, the CPM computational procedure starts by computing the schedule corresponding to point *A* by using all normal activity times and their corresponding normal costs. The procedure then effectively determines where time can be bought the cheapest, and since the activity time-cost curves are assumed to be continuous and linear, the project time-cost curve is made up of a series of straight lines. Thus, all project durations between T_D and T_1 are possible with total direct costs given by the straight line A-A_1. At the latter point, one or more activities have reached their crash times, or a new critical path has been introduced, and time must now be bought on a new activity, or set of activities, which are of necessity at least as expensive (have as great a cost slope) as the previous set. Hence, the next line segment A_1-A_2 has a slope as great as or greater than A-A_1. Finally, when one reaches point B, all activities on at least one of the then existing critical paths have reached their crash times and no further reduction in the project duration is possible.

In the application of the CPM computational procedure, if one is interested in a particular project duration, then one merely computes a portion of the curve A-B until the desired project duration is reached. If, however, one is interested in scheduling the project so as to minimize the total project costs, then one would add the indirect costs to the total direct costs given by the curve A-B and determine the project duration at which the total cost curve reaches a minimum.

It should also be pointed out that the CPM computation procedure can be applied separately to the tasks or subprojects which make up a project. In this case, there will be a total direct-cost curve for each task, similar to the curve A-B in Figure 9-10. Since the latter is continuous, piece-wise linear, and convex, it could, along with similar curves for the other subproject tasks, be used as input data for the development of an over-all project time-cost curve. This procedure is not only useful when the information on the separate tasks is desired along with the over-all project data, but also when the size of the over-all project network exceeds the capacity of the computer being used to process the data. By handling portions of the project network separately and then combining these results, one is able to handle networks of almost any size.

HAND COMPUTATIONAL PROCEDURE BASED ON FEASIBLE ACTIVITY TIME-COST POINTS

The heuristic procedure proposed here is a sequential process consisting of alternate data collection and the rescheduling of activities. Since a reduction in project duration requires a reduction in the duration time of one or more activities on the critical path(s), this process is started by collecting time-cost trade-off data for all critical and near-critical activities. This information is then utilized to reschedule the activities so as to reduce the project duration with a minimum increase in direct costs. In

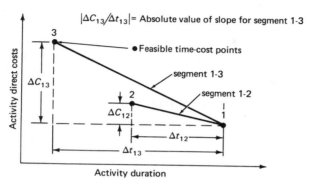

Figure 9-11 Determination of segment slopes.

this rescheduling process, new critical paths will usually be created, which may necessitate the collection of additional activity time-cost trade-off data. This process is stopped when a specified reduction in the project duration has been achieved or when the increase in direct costs exceeds the decrease in indirect costs resulting from the most recent reduction in the project duration.

Feasible time-cost trade-off points are shown in Figure 9-11. No restrictive assumptions about the shape of the time-cost curve are necessary. This particular feature was built into the model in order to provide the necessary flexibility to handle the diversity of situations encountered in practice. Even within the same project, different activities may have cost functions of various shapes, depending on the nature of the work being carried out. It should also be noted that *this model utilizes only the feasible time-cost points,* as shown in Figure 9-11. This feature of the model was adopted because in general real resources are not infinitely divisible, and hence there are only a finite (usually small) number of feasible time-cost points. Another feature of this procedure is that dependencies among the activities with respect to time-cost trade-offs could be stipulated and considered in the hand computational procedure. For example, if a special "resource" is useful in reducing the performance time of a number of activities, then this can be recognized by reducing the performance time of all of these activities simultaneously, if the computational procedure calls for this special resource to reduce the project duration.

Determination of Segment Slopes and Pairwise Slopes

As the next step in the procedure, line segments are drawn physically or conceptually between the (usable) feasible time-cost points, and the slopes of the segments are calculated as shown in Figure 9-11. In real terms, these slopes represent the additional funding required to shorten the activity per unit of time or, alternatively, the cost rate of buying time.

Augmentation Along the Critical Path

The first step in this time-cost trade-off procedure is the determination of a minimum direct-cost schedule, i.e., one which uses only the normal time-cost points for all activities. The determination of the normal project duration is by the conventional forward pass computations described in Chapter 4, and the associated normal project direct costs are obtained merely by summing the normal direct costs for all activities in the project. To reduce the project duration at a minimum increase in project direct costs, the technique of augmenting (increasing) the efforts and direct costs applied to activities along the critical path is utilized. The term augmentation will be used frequently; it refers to an increase in the "efforts" applied to an activity together with a corresponding decrease in its expected duration time.

Augmentations are made beginning with those segments, of critical path activities, having the least (absolute value) slope, which is equivalent to buying time where it can be obtained the cheapest. As each augmentation is made, the forward pass computations are updated to reflect the changes in the length and make-up of the critical path(s) resulting from this activity augmentation.

Time-Cost Trade-off Rules

In the following set of rules, the first one is the basic rule for step-by-step activity segment augmentation, designed to reduce the total project duration with a minimum increase in project direct costs. These rules are not designed to enumerate all of the possible project duration reductions, but they will usually enumerate a sufficient number of points to answer most of the questions of interest. Although following these rules may not give the absolute minimum direct cost increase, they should come very close. Further complexity in the rules, which would in particular be required to handle multiple critical paths optimally, is not felt to be warranted. In certain cases, Rule 1 will lead to Rules 2 or 3 before an activity augmentation is prescribed.

RULE 1. If the network has two or more critical paths, proceed directly to RULE 3. If the network has a single critical path, consider all of the eligible segments of the critical path activities, and augment the segment which has the lowest cost slope, $\Delta c/\Delta t$, choosing the activity with the smallest Δt in case of ties. If this augmentation causes the current critical path to become subcritical, do not make this augmentation, but apply RULE 2 instead.

RULE 2. Let Δt_c denote the smallest reduction in the duration of the current critical path, which just causes one or more additional paths to become critical. Now consider all eligible segments of the current critical path for which $\Delta t \geqq \Delta t_c$, and from among this subset augment that segment which has the lowest cost, Δc, rather than the lowest cost slope, $\Delta c/\Delta t$. Denote the cost slope of the chosen activity segment by $\Delta c_m/\Delta t_m$. Now, if $\Delta t_m = \Delta t_c$, this augmentation cycle is completed; however, if $\Delta t_m > \Delta t_c$, then this augmentation causes the current critical path to go subcritical by an amount equal to say Δt_{sc}, i.e., $\Delta t_m - \Delta t_c = \Delta t_{sc} > 0$. In this case, enumerate as candidates which should be "sold back," those activity segments on this newly formed subcritical path which have been most recently augmented on previous augmentation cycles, and for which $\Delta t \leqq \Delta t_{sc}$. In general, that activity segment, or group of segments, should be sold back, for which the sum of the segment Δt's *is* $\leqq \Delta t_{sc}$, and whose sum of Δc's is a maximum. The resulting cost slope for RULE 2 will then be $(\Delta c_m - \Delta c_{\text{total sold back}})/\Delta t_m$.

RULE 3. Divide the network activities which lie on the critical *paths* into two subsets denoted by I and II. Let subset I contain only those activities common to all critical paths, and subset II those activities not common to all critical paths.

(a) Apply RULE 1, or RULE 2 if required, to subset I to determine the optimal augmentation among these activities. Denote the cost slope of this augmentation by $\Delta c_I/\Delta t_I$.

(b) In order to achieve a project duration reduction by augmenting activities in subset II, two or more activities must be augmented simultaneously. Since no simple set of rules will cover this situation in an optimal manner, it is suggested that likely combinations (based on judgment) of eligible activity segments be augmented to approximate the optimal solution. Let $\Delta c_{II \text{ Net}}$ denote the net augmentation cost which will be equal to the sum of two or more segment augmentations, less the Δc's associated with segments which may be sold back by the principle described in RULE 2. Also let Δt_{II} denote the project duration reduction achieved by the group of augmentations. Then the (approximate) optimal augmentation for subset II activities will have a cost slope equal to $\Delta c_{II \text{ Net}}/\Delta t_{II}$.

(c) Select from the augmentations determined in 3a and 3b, the one with the lowest cost slope.

The illustration of this hand computational procedure will be left as an exercise.

PROCEDURE FOR SCHEDULING TO MEET A SPECIFIED PROJECT DURATION TIME

A useful variation to these computational procedures has been developed by Fondahl,[5] which is appropriate when a project is being planned and scheduled to meet a given total project duration time that obviously requires some augmentation of activities. It is applicable to both the hand computational procedure based on feasible activity time-cost points, as well as the CPM procedure based on the continuous linear time-cost trade-off curves. First, the project is planned and scheduled to meet the specified total project duration time, using crash times *wherever it is felt they are most useful.* Denote this time-cost combination by point S_0 in Figure 9-10. Next, determine whether direct costs can be reduced by taking more time to perform any of the activities located on slack paths, up to the point of their becoming critical. These changes will reduce costs, but will not affect the total project duration time, as indicated by the line S_0-S_1. Finally, the project is alternately lengthened and shortened to "wiggle-in" to the minimum direct cost schedule. The lengthening steps S_1-S_2 and S_3-S_4 are chosen to give the maximum rate of cost decrease, and the shortening steps S_2-S_3 and S_4-S_5 are chosen so as to bring the project back to the specified time at the minimum rate of cost increase. The process is terminated when two successive steps have the same rate of cost change.

SUMMARY

The general philosophy of the time-cost trade-off problem has been presented in this chapter, along with a hand computation procedure based on feasible time-cost points, and the CPM procedure based on continuous linear time-cost curves and a rigorous computation algorithm.

The CPM computational procedure is, indeed, a most powerful tool; when its assumptions are satisfied, its use is more economical than the hand computational procedure previously described, even if the size of the network is quite small. For this reason, it may be possible in some cases to exploit its efficiency, even where the assumptions regarding the individual activity time-cost curves are not all satisfied. This could be accomplished by first changing the time-cost curves, where necessary, so that they are continuous, piece-wise linear, decreasing, and convex; one may assume they are all independent, and then obtain the CPM solution to the problem. Having this solution as a starting point, one then may be able to alter it by hand, taking into consideration the important deviations from the assumptions.

While the above procedure has merit, it has the serious shortcoming of not providing an estimate of how far the final schedule may be removed from the optimal schedule. If the magnitude of the costs involved warrants a more refined treatment of this problem, the linear (integer) programming methods developed by Meyer and Shaffer,[6] presented in Appendix 9-1, should be considered. These methods are quite flexible in that they will treat a completely general type of activity time-cost trade-off relationship, and they can be extended, for example, to treat certain types of trade-off dependencies among the project activities. Unfortunately, the solution of problems formulated in this manner requires the use of large computing machines, and is presently restricted to networks of about 50 activities. The latter difficulty can be overcome, however, by analyzing portions (subprojects) of the entire project network, and then analyzing these results as though they were single activities to obtain a solution for the entire project.

The time-cost trade-off problem is similar to the problem of scheduling activities to satisfy resource constraints, treated in Chapter 8, in that a wide range of techniques is available. If a project network is available, the principles studied in this chapter can be applied in a very informal way, without the aid of computers or involved algorithms, to make intelligent choices of activity augmentations to reduce the project duration by specified amounts. If the objective is to minimize the total project costs, direct plus indirect, then more formal procedures, such as those presented in detail in this chapter and its appendix, are useful.

REFERENCES

1. Kelley, J. E., and M. R. Walker, *Critical Path Planning and Scheduling*, Proceedings of the Eastern Join Computer Conference, December, 1959, pp. 160-173.

2. Kelley, J. E., Jr., "Critical Path Planning and Scheduling: Mathematical Basis," *Operations Research,* Vol. 9, No. 3 (1961), pp. 296-320.
3. Prager, W., "A Structural Method of Computing Project Cost Polygons," *Management Science,* Vol. 9, No. 3 April (1963), pp. 394-404.
4. Fulkerson, D. R., "A Network Flow Computation for Project Cost Curves," *Management Science,* Vol. 7, No. 2, January (1961), pp. 167-179.
5. Fondahl, J. W., "Can Contractors Own Personnel Apply CPM Without Computers," *The Constructor,* November (1961) pp. 56-60, and December (1961) pp. 30-35.
6. Meyer, W. L., and L. R. Shaffer, *Extensions of the Critical Path Method Through the Application of Integer Programming,* Report issued by the Dept. of Civil Engineering, University of Illinois, Urbana, Ill., July 1963.
7. Charnes, A., and W. W. Cooper, "A Network Interpretation and a Directed Subdual Algorithm for Critical Path Scheduling," *Journal of Industrial Engineering,* Vol. 13, No. 4 (1962) pp. 213-218.
8. Ford, L. R., and D. R. Fulkerson, *Flows in Networks,* Princeton University Press, Princeton, N. J., 1962.
9. Hadley, G., *Linear Programming,* Addison-Wesley Publishing Company, Reading, Massachusetts, 1962.

EXERCISES

1. Given below are the network data and the time-cost trade-off data for a small maintenance project.

Table 9-2.

Job (Activity)	Predecessor Jobs	Normal (days)	Normal (dollars)	Crash (days)	Cost Slope (dollars/day)
A	none	3	50	2	50
B	none	6	140	4	60
C	none	2	50	1	30
D	A	5	100	3	40
E	C	2	55	2	—
F	A	7	115	5	30
G	B, D	4	100	2	70
Total			610		

Assume the indirect costs, including the cost of lost production, and associated outage costs, supervision, etc., to be as follows.

Project duration (days)	12	11	10	9	8	7
Indirect costs (dollars)	900	820	740	700	660	620

Assuming any integer times between the normal and crash activity times are feasible, use the hand computational time-cost trade-off procedure to show

Table 9-3.

Activity No.	Predecessor Activity Nos.	Description of Activity	Time (hours)			Cost (dollars)	
			a	m	b	Normal	Crash
101	—	inspect & measure pipe	2	4	5	16	22
102	—	devlp. cal. mtls. list	3	6	8	18	25
103	101	make drawings of pipe	2	3	5	12	18
104	—	deactivate line	7	8	10	8	14
105	102	procure calender parts	120	244	320	12	35
106	102	assemble calender work crew	6	8	9	20	30
107	103	devlp. matl. list (pipe)	3	4	7	10	13
108	104, 105, 106	deactivate calender	4	4	5	3	3
109	107	procure valves	136	220	280	10	20
110	107	procure pipe	136	200	240	10	22
111	107	assemble work crew (pipe)	4	6	7	16	20
112	108	tie off warps	1	2	3	3	8
113	110, 111	prefab pipe sections	20	40	50	120	240
114	111	erect scaffold	6	12	15	30	65
115	112	disassemble calender	4	10	14	90	210
116	112	empty & scour vats	2	3	5	6	9
117	104, 114	remove old pipe	18	30	38	180	300
118	115	repair calender	35	70	98	650	1500
119	113, 117	position new pipe	6	8	12	50	110
120	118	lubricate calender	3	5	6	10	22
121	109, 119	position new valves	5	7	10	66	100
122	119	weld new pipe	6	8	11	50	60
123	120	reassemble calender	18	22	24	200	270
124	123	adjust & balance calender	6	8	14	80	95
125	121, 122	insulate pipes	15	20	30	60	75
126	121, 122	connect pipes to boiler	3	4	5	24	30
127	121, 122, 123	connect pipes to calender	7	8	11	48	50
128	116, 124	refill vats	1	1	1	2	2
129	125, 126	remove scaffold	3	4	4	16	18
130	126	pressure test	5	6	9	15	16
131	128	tie in warps	3	4	5	8	10
132	127, 131	activate calender	2	2	3	14	14
133	129, 130, 132	clean up	3	4	5	15	18

that the total costs are $1510, 1470, 1430, 1470, 1530 and 1620 for 12, 11, 10, 9, 8, and 7 day project durations, respectively.

2. Repeat exercise 1 assuming each activity has a continuous linear time-cost trade-off curve, and using CPM computational algorithm in Appendix 9–1.

3. Given in Table 9-3 are the data for a steam calender and pipeline maintenance project. Three sets of times are given under columns headed *a, m,* and *b*. In this problem only the first two columns will be used: the times

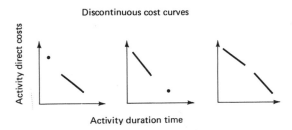

Figure 9-12

under the column headed a are the "crash" activity performance times while those under the column headed m are the "normal" performance times.

a. Draw the network and make the basic scheduling computations using activities-on-arrows or activities-on-nodes.

b. Identify the critical path.

c. Indicate, in step-by-step detail, the activity augmentations required to reduce the project duration to 248 hours while keeping the total project direct costs at a minimum. Assume that only the normal and crash times are feasible.

d. Repeat part c using the method illustrated in Figure 9-10 by the points $S_0, S_1, \ldots, S_5$.

4. Write out a detailed set of instructions for the third hand computational time-cost trade-off rule, which could be programmed on a computer.

5.†In the appendix to this chapter, the linear programming formulations are given for activity time-cost trade-off curves that are either continuous, or a collection of feasible time-cost points. These formulations can be extended to cover the case where combinations of these two types of trade-off functions occur. An analysis of this problem indicates that the transitions can occur in three distinct ways as shown in Figure 9-12, i.e., any combinations of continuous curves and discrete points can be handled by appropriate combinations of these three transitions. Write out the linear programming formulation for each of these transitions.

6. An application of the CPM time-cost trade-off procedure is given on pages 232-235 of Appendix 9-1. As shown there, it is a "tabular" node labeling procedure. Develop a scheme for carrying out this algorithm on the network itself. Suggestion: draw and label the network as shown in Fig. 9-14, adding a space for the current flow f_{ij1}, and add the activities that carry the flow f_{ij2}, only as needed.

7.†Consider the simple project network shown in Figure 9-13, along with the accompanying activity time-cost trade-off curves. Using the linear (integer)

---

† This problem pertains to material taken up in the appendix to this chapter.

programming techniques described in the appendix to this chapter, write the linear programming formulation of this problem, and determine the schedule of activity duration times (y_{ij}) which will minimize the total direct project costs for a project duration constraint of $\lambda = 10$ time units.

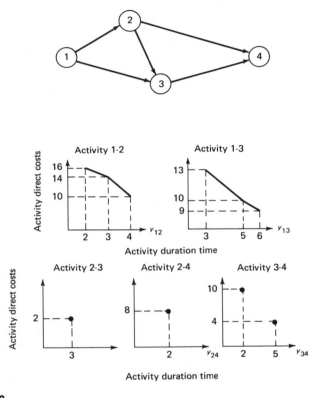

Figure 9-13

APPENDIX 9-1—APPLICATIONS OF LINEAR PROGRAMMING TO THE DEVELOPMENT OF PROJECT COST CURVES

The purpose of this appendix is to present the current applications of linear programming techniques to the problem of generating project cost curves. The treatment given here will assume that the reader is familiar with the basic linear programming formulation of an optimization problem. However, *network flow* theory will be covered in somewhat greater detail, since this is generally not so well known.

First, the basic network scheduling (forward pass) computations will be viewed as a problem in linear programming; this approach was published by Charnes and Cooper.[7] The time-cost trade-off problem will then be introduced, and the integer-programming methods developed by Meyer and Shaffer[6] will be presented to handle very general types of activity time-cost trade-off functions. Finally, general network flow theory will be introduced, together with its application to the project network time-cost trade-off problem, as developed in a basic paper by D. R. Fulkerson,[4] and a text titled "Flows in Networks" by L. R. Ford and D. R. Fulkerson.[8]

LINEAR PROGRAMMING FORMULATION OF BASIC SCHEDULING COMPUTATIONS

The various topics treated in this appendix will be illustrated using the simple network† shown in Figure 9-14. The numbers appearing along each activity in this figure give (d_{ij}, D_{ij}, C_{ij}). For example, (d_{12}, D_{12}, C_{12}) = (1,3,3) indicates that for activity 1-2, the crash and normal performance times are 1 and 3 time units, respectively, and the slope of the linear time-cost trade-off curve is 3 monetary units per time unit. In the applications, the time unit is chosen so that the d_{ij}'s and D_{ij}'s are all integer valued, and the monetary unit is chosen so that the C_{ij}'s are also integer valued.

For purposes of illustrating how the basic forward pass computations can be formulated as a linear programming problem, one may assume that all activities in Figure 9-14 are scheduled to be performed at their normal times, i.e., the middle number of the three numbers given for each activity. Now the project network may be viewed as a *flow network*, in which a hypothetical unit of flow leaves the source, node (event) 1, and enters the sink, node 4. Also, nodes 2 and 3 play the role of "trans-

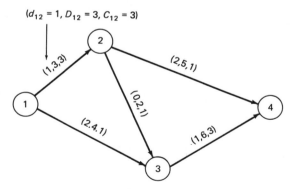

$(d_{12} = 1, D_{12} = 3, C_{12} = 3)$

Figure 9-14 Illustrative network with data giving crash time, normal time, and cost slope for each activity.

† By permission from L. R. Ford, Jr., and D. R. Fulkerson, "Flows in Network," The Rand Corporation, 1962. Published, 1962, by Princeton University Press.

shipment" points and, thus, at these nodes there must be a *conservation of flow*, i.e., the total flow into node 2 must equal the total flow away from node 2, and similarly for node 3. The performance time of each activity, y_{ij}, is then interpreted as the time (or cost) of transporting *a unit of flow* from node i to node j. Viewing a project network in this way reduces the problem of finding the critical path(s) to the determination of the network path(s) or route(s) from the source, node 1, to the sink, node 4, which requires a maximum time (or cost) to traverse. The reader is no doubt well aware of the fact that there are easier ways of locating the network critical path. The main purpose of this *network flow* interpretation of the problem is to show that it can be formulated as a linear programming problem, a technique which will prove to be useful later in solving problems which cannot be easily solved in other ways.

Applying the above *network flow* interpretation to Figure 9-14 results in the linear programming formulation given in equation (1), which will be referred to as the *primal* problem. In this formulation, $y_{ij} = 0$ or 1 denotes the absence or presence of a unit flow from node i to node j, a flow which is said to be along activity $i\text{-}j$.

PRIMAL PROBLEM

Maximize $f[\mathbf{Y}] = 3y_{12} + 4y_{13} + 2y_{23} + 5y_{24} + 6y_{34}$ (1a)

$$
\begin{aligned}
\text{Subject to} \quad && y_{12} + \ y_{13} && && && &= 1 \quad &&(1b) \\
&& -\,y_{12} && +\ y_{23} +\ y_{24} && && &= 0 \quad &&(1c) \\
&& -\,y_{13} -\ y_{23} && && +\ y_{34} &= 0 \quad &&(1d) \\
&& && -\ y_{24} -\ y_{34} && &= -1 \quad &&(1e)
\end{aligned}
$$

$$y_{12} \geqq 0,\ y_{13} \geqq 0,\ y_{23} \geqq 0,\ y_{24} \geqq 0,\ y_{34} \geqq 0$$

Constraint equations (1b) and (1e) indicate that a unit of flow leaves the source, node 1, and enters the sink, node 4, while constraint equations (1c) and (1d) require a conservation of flow at the intermediate nodes 2 and 3. Thus., any set of y_{ij}'s which satisfy these constraints constitute a path from the source to the sink, as indicated by the y_{ij}'s that are equal to one. Then, since the y_{ij}'s are equal to one for the activities carrying the hypothetical unit flow, and zero otherwise, the objective function, $f[\mathbf{Y}]$, gives the sum of the activity duration times for the chosen path from source to sink. When the objective function is maximized, the corresponding path, denoted by $\mathbf{Y}^*$, is the longest or critical path through the network; the corresponding value of the objective function will be denoted by $f[\mathbf{Y}^*]$. The solution of this problem is greatly facilitated by the use of the duality theorem of linear programming as described below.

Formulation of the Dual Problem

From the well known duality theorem of linear programming, it can be shown that to every linear programming problem, call it the *primal,* there corresponds another linear programming problem which is called the *dual* of the original problem. The connection between these two problems is stated in the following theorem.

Duality Theorem

Given a linear programming problem, call it the primal, there is always a related linear programming problem, called the dual, defined as follows, in matrix notation.

PRIMAL

$$\text{Maximize } f[\,\mathbf{X}\,] = \underset{1 \times n}{[\,\mathbf{C}\,]} \underset{n \times 1}{[\,\mathbf{X}\,]}$$

$$\text{Subject to } \underset{m \times n}{[\,\mathbf{A}\,]} \underset{n \times 1}{[\,\mathbf{X}\,]} \leqq \underset{m \times 1}{[\,\mathbf{P}\,]}$$

$$\underset{n \times 1}{[\,\mathbf{X}\,]} \geqq \underset{n \times 1}{[\,\mathbf{O}\,]}$$

DUAL

$$\text{Minimize } g[\,\mathbf{W}\,] = \underset{1 \times m}{[\,\mathbf{P}'\,]} \underset{m \times 1}{[\,\mathbf{W}\,]}$$

$$\text{Subject to } \underset{n \times m}{[\,\mathbf{A}'\,]} \underset{m \times 1}{[\,\mathbf{W}\,]} \geqq \underset{n \times 1}{[\,\mathbf{C}\,]}$$

$$\underset{m \times 1}{[\,\mathbf{W}\,]} \geqq \underset{m \times 1}{[\,\mathbf{O}\,]}$$

If there exists a solution, $[\mathbf{X}^*]$, which gives a finite maximum value to
$\quad\quad\quad\quad\quad n \times 1$
$f[\mathbf{X}]$, there is always a "coupled" solution $[\mathbf{W}^*]$, for which $g[\mathbf{W}]$ has
$\quad\quad\quad\quad\quad\quad\quad\quad m \times 1$
a finite minimum value, equal to $f[\mathbf{X}]$. (A solution to one problem makes the solution to the other problem readily available.)

To the above statement of the theorem must be added that while the primal problem has n variables and m constraints, the dual problem has the reverse, m variables and n constraints. Also, if any of the n primal variables are unrestricted in sign, then the corresponding dual constraints are equalities, and if the primal constraints are equalities, then the dual variables are unconstrained in sign. With regard to the solutions of the primal and dual problems, if in the solution of the dual one of its inequality constraints is satisfied as an equality, then the primal variable corresponding to this dual constraint may be positive, whereas it must be zero if the dual constraint is satisfied as an inequality.

Applying the Duality Theorem to the above example, one obtains the dual formulation given in equation (2). Recall that in the primal there were 5 variables and 4 constraints; hence in the dual problem there are

5 constraints (one corresponding to each of the primal variables) and 4 variables (one corresponding to each of the primal constraints.) Also, since the primal constraints were all equalities, the dual variables are all unconstrained in sign.

DUAL PROBLEM

Minimize $g[\mathbf{W}] = w_1 - w_4$ (2a)

$$
\begin{array}{llll}
\text{Subject to } w_1 - w_2 & & \geqq 3 & \text{(2b)} \\
w_1 & - w_3 & \geqq 4 & \text{(2c)} \\
w_2 & - w_3 & \geqq 2 & \text{(2d)} \\
w_2 & - w_4 & \geqq 5 & \text{(2e)} \\
& w_3 - w_4 & \geqq 6 & \text{(2f)}
\end{array}
$$

$$ -\infty < w_1 < \infty, \; -\infty < w_2 < \infty, \; -\infty < w_3 < \infty, \; -\infty < w_4 < \infty $$

The advantage of the dual formulation to this problem is now fairly obvious. Since each dual constraint involves only two variables, they can be solved by inspection, if a value is assigned to w_1. To see this, consider the constraints written in the following equivalent form.

$$
\begin{array}{lll}
w_2 \leqq w_1 - 3 & w^{*}_2 = -3 & \text{(2bb)} \\
w_3 \leqq w_1 - 4 \atop w_3 \leqq w_2 - 2 & w^{*}_3 = -6 & {\text{(2cc)} \atop \text{(2dd)}} \\
w_4 \leqq w_2 - 5 \atop w_4 \leqq w_3 - 6 & w^{*}_4 = -12 & {\text{(2ee)} \atop \text{(2ff)}}
\end{array}
$$

It will be shown below that w_4 varies directly with the value assigned to w_1, and since $g[\mathbf{W}]$ is merely the difference between w_1, and w_4, w_1 can be assigned an arbitrary value without affecting $g[\mathbf{W}]$. Thus, one may let $w_1 = 0$, a choice which makes all of the other w_i's, and in particular w_4, take on negative values. Then, to minimize $g[\mathbf{W}] = w_1 - w_4 \equiv -w_4$, one must minimize the absolute value of w_4, and at the same time satisfy each of the above constraints. It is easy to see by inspection that this is achieved by the solution indicated above, i.e., $w^{*}_1 = 0$, $w^{*}_2 = -3$, $w^{*}_3 = -5$ and $w^{*}_4 = -11$. This will be denoted as $[\mathbf{W}^{*}] = (0, -3, -5, -11)$. Thus, the optimal value of the dual objective function is $g[\mathbf{W}^{*}] = w^{*}_1 - w^{*}_4 = 0 - (-11) = 11$. It is interesting to note the similarity between the solution of inequalities (2bb) through (2ff), and the conventional forward pass computations. For example, the value of $w^{*}_3 = -5$ is the largest absolute value of the two solutions obtained by treating (2cc) and (2dd) as equalities, or equivalently, the smallest absolute value of w_3 which satisfies both of these constraints. This is analogous to calculating the earliest event time at a merge point by taking the largest of the earliest finish times of the merging activities, which are 4 and 5 in this case.

Noting that the second and fourth dual constraints, i.e., those given by equations (2c) and (2e) are satisfied as inequalities, one knows from the duality theorem that the second and fourth primal variables, i.e., y_{13} and y_{24} are zero, while the others may be positive. By checking the primal constraints equations (1b) through (1e), one sees that the optimal solution to the primal problem is given by $[\mathbf{Y}^*] = (y^*_{12}, y^*_{13}, y^*_{23}, y^*_{24}, y^*_{34}) = (1,0,1,0,1)$, and that $f[\mathbf{Y}^*] = 3 \times 1 + 4 \times 0 + 2 \times 1 + 5 \times 0 + 6 \times 1 = 11$. (Note also that the requirement, $f[\mathbf{Y}^*] = g[\mathbf{W}^*] = 11$ is satisfied.) Thus, it is concluded that the network critical path is comprised of activities 1-2, 2-3, and 3-4, and has a total duration of 11 time units, a value which can easily be verified by the routine forward pass computation.

LINEAR PROGRAMMING FORMULATION OF TIME-COST TRADE-OFF PROBLEM

The basic formulation of the time-cost trade-off problem will be given first, assuming each activity in the network shown in Figure 9-14 has a trade-off curve of the form shown in Figure 9-15. The advantage of the notation used in this figure is that it replaces the normal and crash costs by a single cost slope, C_{ij}. Again letting y_{ij} denote the scheduled duration of activity i-j, one can write the total direct project costs as a function of these variables as follows:

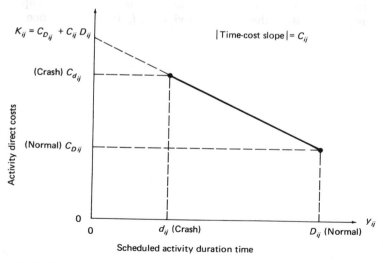

Figure 9-15 Time-cost trade-off curve nomenclature.

$$\text{Total direct project costs} = \sum_i \sum_j (K_{ij} - C_{ij}y_{ij}) = K - \sum_i \sum_j C_{ij}y_{ij} \quad (3)$$

where $\sum_i \sum_j$ is used to denote the summation over all activities in the

project network. Since the K_{ij}'s are fixed constants, whose sum is denoted by K in equation (3), the total direct project costs are minimized if one will*

* In this appendix, the symbol T_i will be used in place of E_i, used elsewhere in this text.

$$\text{Maximize } f[\mathbf{Y}] = \sum_i \sum_j C_{ij}y_{ij} \quad (4a)$$

$$\text{Subject to } T_i + y_{ij} - T_j \leq 0, \text{ all } ij \quad (4b)$$

$$y_{ij} \leq D_{ij}, \text{ all } ij \quad (4c)$$

$$- y_{ij} \leq - d_{ij}, \text{ all } ij \quad (4d)$$

$$T_4 - T_1 \leq \lambda \quad (4e)$$

where the T_k's are (unknown) variables denoting the earliest expected time for node $k(k = 1, 2, 3, \text{ or } 4)$, and λ is the (constant) constraint placed on the total project duration, which is merely $T_4 - T_1$. Following the CPM convention of letting $T_1 = 0$, the last constraint, equation (4e) would merely become $T_4 \leq \lambda$.

The first constraint equation (4b) applies to each activity, i-j, in the network, of which there are five in this example. These constraints merely state that for activity i-j, the difference between the earliest node times, T_i and T_j, must be at least as great as y_{ij}, the scheduled duration of activity i-j. Similarly, equation (4c) applies to each of the five activities in this example; it constrains the scheduled activity duration time, y_{ij}, to be equal to or less than the normal activity time, D_{ij}. Finally, equation (4d) constrains each y_{ij} to be equal to or greater than the crash activity time, d_{ij}. The variables, T_k, have been omitted from the objective function because their cost coefficients are zero; their role in this formulation is merely to insure that the scheduled values of y_{ij} are feasible from the standpoint of network logic, and to insure that the project duration does not exceed λ. This linear programming formulation, as stated in equations (4a) through (4e), has been written out in full in Table 9-4. While this problem could be solved using the simplex method to find the schedule $[\mathbf{Y}^*] = (y^*_{12}, y^*_{13}, y^*_{23}, y^*_{24}, y^*_{34})$ which satisfies all of the constraints and at the same time maximizes $f[\mathbf{Y}]$, a more efficient network flow algorithm will be given at the end of this appendix.

Formulation for Continuous Convex Activity Time-Cost Trade-off Curves

Chapter 9 discussed the use of a piece-wise linear approximation of a nonlinear but convex activity time-cost trade-off function. Suppose, for example, the trade-off curve for activity i-j is as shown in Figure 9-16 below, which is continuous, convex, and nonincreasing. The actual trade-off curve is shown at the left of Figure 9-16, with the two curves on the right depicting the separate segments of the piece-wise linear approximation to the actual curve. This representation of y_{ij} in the objective function and the constraint equations is given below, where only that portion of the formulation pertaining to activity i-j is given.

$$\text{Maximize } f[Y] = \ldots + C_1 y_{1ij} + C_2 y_{2ij} + \ldots \qquad (5a)$$

Subject to

$$\begin{aligned}
T_i + d_{ij} + y_{1ij} + y_{2ij} - T_j &\leq 0 & (5b)\\
0 \leq y_{1ij} &\leq m - d & (5c)\\
0 \leq y_{2ij} &\leq D - m & (5d)
\end{aligned}$$

Using the simplex method to solve this problem will bring y_{1ij} and y_{2ij} into the solution in the proper order, i.e., y_{1ij} will remain its maximum value until y_{2ij} is reduced to zero, and then only will y_{1ij} be reduced below its maximum value. This follows because $C_1 > C_2$, and the sum $C_1 y_{1ij} + C_2 y_{2ij}$ is being maximized. It is obvious that this would prevail for any number of straight line segments, as long as the actual trade-off is convex, which insures that $C_i > C_{i+1}$, for all i.

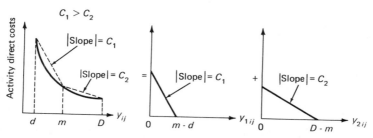

Figure 9-16 Breakdown of a continuous convex activity time-cost trade-off curve.

Table 9-4.   Primal Linear Programming Formulation of The Time-Cost Trade-off Problem

Maximize $C = \sum\limits_{i=1}^{4} C_i T_i + \sum\limits_{i}\sum\limits_{j} C_{ij} y_{ij}$

$\quad\quad = 0T_1 + 0T_2 + 0T_3 + 0T_4 + C_{12} y_{12} + C_{13} y_{13} + C_{23} y_{23} +$
$\quad\quad\quad\quad C_{24} y_{24} + C_{34} y_{34}$

Subject to

T_1	T_2	T_3	T_4	y_{12}	y_{13}	y_{23}	y_{24}	Y_{34}	$\leqq$	Constraint
1	−1			1					$\leqq$	0
1		−1			1				$\leqq$	0
	1	−1				1			$\leqq$	0
	1		−1				1		$\leqq$	0
		1	−1					1	$\leqq$	0
				1					$\leqq$	D_{12}
					1				$\leqq$	D_{13}
						1			$\leqq$	D_{23}
							1		$\leqq$	D_{24}
								1	$\leqq$	D_{34}
				−1					$\leqq$	$-d_{12}$
					−1				$\leqq$	$-d_{13}$
						−1			$\leqq$	$-d_{23}$
							−1		$\leqq$	$-d_{24}$
								−1	$\leqq$	$-d_{34}$
−1		1							$\leqq$	λ

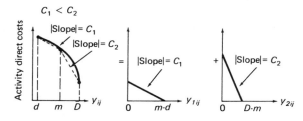

Figure 9-17   Breakdown of a continuous concave activity time-cost trade-off curve.

Formulation for Continuous Nonconvex Activity Time-Cost Trade-off Curves

If the trade-off curve is continuous, concave, and nonincreasing, the segments of a piece-wise linear approximation to the actual curve can be brought into the problem in their proper order by employing a non-negative integer-valued variable. An approximation of this type is shown in Figure 9-17 above, where the actual curve is shown on the left and the separate segments of this piece-wise linear approximation are shown on the right. In this case, the representation of y_{ij} in the objective function and the constraint equations is given below, where again only that portion of the formulation pertaining to activity i-j is given

$$\text{Maximize } f[\mathbf{Y}] = \ldots + C_1 y_{1ij} + C_2 y_{2ij} + \ldots \tag{6a}$$

Subject to

$$T_i + d_{ij} + y_{1ij} + y_{2ij} - T_j \leq 0 \tag{6b}$$
$$0 \leq y_{1ij} \leq m - d \tag{6c}$$
$$0 \leq y_{2ij} \leq D - m \tag{6d}$$
$$\delta(m - d) \leq y_{1ij} \tag{6e}$$
$$\delta(D - m) \geq y_{2ij} \tag{6f}$$
$$\delta = \text{a non-negative integer} \tag{6g}$$

If one puts constraints (6e) and (6f) together, one finds that this system of constraints actually requires that δ be equal to either zero or one, as shown in equation (7).

$$y_{2ij}/(D - m) \leq \delta \leq y_{1ij}/(m - d) \tag{7}$$

If y_{ij} is equal to its maximum value, then $\delta = 1$ since $y_{2ij} = D - m$, and because of the integer constraint on δ, it must continue to equal 1 as long as $y_{2ij} > 0$. Consequently, y_{1ij} must, as it should, remain equal to its maximum value of $m - d$, as long as $y_{2ij} > 0$. When $y_{2ij} = 0$, δ can also equal zero, and only then can y_{1ij} be less than $m - d$. Thus, one can see that the use of the integer-valued variable, δ, forces the activity segments to vary in a manner dictated by the physical problem that they represent.

It is a simple matter to extend this formulation to more than two linear segments by introducing an additional integer-valued variable for each additional segment as shown in equation (8).

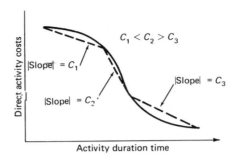

Figure 9-18   Illustrative nonconvex, nonconcave trade-off function.

$$y_{k+1}/(d_{k+1} - d_k) \leqq \delta_k \leqq y_k/(d_k - d_{k-1}); k = 1, 2, \ldots, n \quad (8)$$

where in general the constraint, $0 \leq y_i \leq d_i - d_{i-1}$, holds. Also, it is important to note that this formulation of the problem will work on convex trade-off function as well as concave functions; however, it is not needed in the former case, as shown by equation (5). Because this formulation can be extended to any number of straight line segments, concave or convex, it follows that this integer-variable formulation can be used on any trade-off curve that is continuous and nonincreasing, such as the curve shown in Figure 9-18 which is neither convex nor concave.

Formulation for Feasible Point Time-Cost Trade-off Functions

It often happens that an activity can only be performed in a small number of different times, which gives rise to a small set of feasible time-cost points. For example, consider an activity *i-j* that can only be performed at a normal time, D_{ij}, or a crash time, d_{ij}. In this formulation of the problem, one requires two† non-negative integer-valued variables as follows:

$$\begin{aligned} y_{D_{ij}} &= 1 \text{ if the activity duration is } D_{ij} \\ &= 0 \text{ if the activity duration is } d_{ij} \\ y_{d_{ij}} &= 1 \text{ if the activity duration is } d_{ij} \\ &= 0 \text{ if the activity duration is } D_{ij} \end{aligned} \quad (9)$$

Using these integer-valued variables, the linear programming formulation is as follows:

---

† One variable would suffice in this case, since $y_{D_{ij}} = 1 - y_{d_{ij}}$; however, the formulation given in equation (10) is used because it suggests the generalization to more than two feasible time-cost points.

Maximize $f[\mathbf{Y}] = \cdots - C_{D_{ij}} y_{D_{ij}} - C_{d_{ij}} y_{d_{ij}} + \cdots$ (10a)
Subject to

$$\begin{aligned} &\cdot \\ &\cdot \\ &\cdot \end{aligned}$$

$$T_i + D_{ij} y_{D_{ij}} + d_{ij} y_{d_{ij}} - T_j \leqq 0 \qquad (10b)$$
$$y_{D_{ij}} + y_{d_{ij}} = 1 \qquad (10c)$$
$$y_{D_{ij}}, \ y_{d_{ij}} = \text{non-negative integers} \qquad (10d)$$

$$\begin{aligned} &\cdot \\ &\cdot \\ &\cdot \end{aligned}$$

If the activity has k different feasible time-cost points, the above formulation is extended by introducing one non-negative integer-valued variable for each feasible time-cost point and requiring that the sum of all of the variables be equal to one.

The use of integer variables in this manner is, indeed, a very powerful tool which could be used in a number of other ways. Consider, for example, the criticism put forth by Fondahl[5] that the CPM procedure does not account for the fact that activities are sometimes correlated in that a speed up of activity A must be accompanied by a speed up in say activity B as well, because it is accomplished by the use of a special resource which acts in common to both of these activities. In this case, one could accomplish the requirement that neither or both of the activities A and B are augmented by adding a pair of constraints to equation (10), i.e., $y_{d_A} - y_{d_B} \leqq 0$ and $y_{d_B} - y_{d_A} \leqq 0$, where y_{d_A} and y_{d_B} are already required to be non-negative integers by constraints like (10d). This pair of constraints accomplishes the desired result because they require, in effect, that both variables must be equal to zero or both must be equal to one.

Solving Problems with Arbitrary Trade-off Functions

It is not too difficult to combine the above formulations to handle more complicated trade-off functions which are combinations of discrete points and continuous curves. The details of this procedure can be found in the report by Meyer and Shaffer;[6] the three basic transition forms that must be handled are enumerated in problem 5 at the end of Chapter 9.

The application of the linear programming formulations given by equations (4), (5), (6), and (10), to the network given in Figure 9-14 but with more complicated activity trade-off functions, is presented in problem 6. The solution of this very small (five activities) problem requires 13 constraints, and the restriction that two of the variables (one for the concave cost curve for activity 1-2, and one for the discrete cost points for activity 3-4) be integer valued. To solve even this small problem using the simplex method, together with an integer programming algorithm, will require the use of a computer if the task is to be accomplished in a reasonable time and at a reasonable cost. Meyer and Shaffer estimate that one of the largest present-day computers, together with current computer programs, is not capable of handling networks containing much more than 50 activities. Larger networks can be handled by treating small portions of the network as separate problems, and then piecing these solutions together to obtain an over-all solution. This procedure, however, is not a simple task. Although improvements in computers, and particularly in computer programs to solve this type of problem can be expected, the use of linear programming in solving time-cost trade-off problems will be restricted to very costly projects unless simpler methods of formulating and solving the problem are made possible. Fortunately, a great simplification is possible, if all of the activity time-cost trade-off curves are continuous, convex, and nonincreasing as shown in Figure 9-16. Under these assumptions the problem can be formulated as a *network flow* problem for which an extremely efficient algorithm is available; it can handle networks with thousands of activities even on moderate sized computers. The next section will treat the general problem of *flows in networks,* and then this appendix will be concluded by the development of the *network flow* algorithm to solve the time-cost trade-off problem.

NETWORK FLOW THEORY

This section will be based on the following set of definitions, which are presented in a more complete manner by Hadley.[9]

Graph

A set of two or more nodes with certain pairs of these nodes joined by one or more lines (called branches) will be called a graph.

In applications to critical path methods, the nodes are called events and the branches are called activities. The nodes will be numbered, and they in turn will identify the individual branches.

Path

A path joining nodes i and j is an ordered set of branches, (i,p), (p,q), . . . (t,u), (u,j), such that each node in the ordered set, with the possible exception of the first and last nodes, is the end point for two and only two branches in the set. The nodes i and j are called the extremity points of the path.

Loop

If a path is defined so that $i = j$, then the path is called a loop.

Network

A network is a graph such that a flow can take place in the branches of the graph.

A sense of direction can be attributed to a branch by stating which node is to be considered the point of origin. Such a branch will be called *oriented*. For example, $(i \rightarrow j)$ denotes a branch originating at node i and terminating at node j.

Oriented Network

An oriented network is one whose branches are all oriented. Similarly, a path, or loop, which follows the network branch orientation will be referred to as oriented.

If a network is oriented, the orientation of any branch is taken to be the direction of flow, which may be limited by a capacity restriction on the branch flow. The flow may be an electric current, a fluid, goods, money, etc.

A node u is called a *source* if every oriented branch which has this node as an end point is oriented so that the flow in these branches moves away from node u to another node. Similarly, a node v is called a *sink* if all oriented branches touching node v carry flow into v.

Project Network

A project network is an oriented network that has an oriented path from a single source node to every other node in the network, and from every node in the network to a single sink node, but contains no oriented loops.

Maximal Flows in Networks—Linear Programming Formulation

Consider a project network whose branches have capacity restrictions denoted by $d_{ij} \geqq 0$, which may be finite or infinite. The problem is to compute the maximum possible flow (steady-state flow rate) from source to sink. If the flow from node i to j is denoted by y_{ij}, then it must satisfy the constraint $0 \leqq y_{ij} \leqq d_{ij}$, and the flow in this branch can be increased by any amount up to a maximum value equal to $e_{ij} = d_{ij} - y_{ij}$; the e_{ij}'s will be referred to as excess capacities. The linear programming formula-

tion can then be written as follows, where the source node is numbered
1 and the sink node, N.

$$\text{Maximize } f[\mathbf{Y}] = \sum_r y_{1r} = \sum_s y_{sN} \qquad (11\text{a})$$

$$\text{Subject to } \sum_s y_{si} = \sum_r y_{ir}; \ i = 2, \ldots, N-1 \qquad (11\text{b})$$

$$0 \leq y_{ij} \leq d_{ij}; \text{ all i,j} \qquad (11\text{c})$$

The summations used here are understood to be over all branches in the
network. Equation (11a), which is to be maximized, gives the total network flow, which is equal to the total flow leaving the source node 1, and
the flow entering the sink node N. Constraint equation (11b) requires
conservation of flow at all intermediate network nodes, and constraint
equation (11c) requires that all scheduled flows stay within their prescribed limits. This problem could, of course, be solved by the simplex
method; however, a much simpler flow algorithm shall be developed.

Intuitive Maximal Network Flow Algorithm

The following intuitive approach to solving this problem is given below,
and will be applied to the illustrative network shown in Figure 9-19a.

STEP 1. Starting at the source, move along branches of positive capacity
until you reach the sink. Enumerate the capacities of these
branches, denote the minimum of these capacities by d, set the
flow in the path equal to d, and finally deduct d from each of
the branch capacities along the path in question to obtain the
remaining or excess branch capacity.

STEP 2. Repeat the above operation for the new network capacities until
there is no longer a path from source to sink with positive capacities. The maximal flow is then the sum of the flows in the paths
which were obtained at each step. In carrying out STEP 2, a flow
in the wrong direction (against the branch orientation) is permissible as long as the net flow is nonnegative. If one lets y'_{ij}
and y'_{ji} denote simultaneous flows in the same branch but in
opposite directions, then the net flow y_{ij} is given by equation (12).

$$d_{ij} \geq y_{ij} = y'_{ij} - y'_{ji} \geq 0, \text{ and } y'_{ji} \leq y'_{ij} \qquad (12)$$

Thus, one can compute the excess capacities for branch $i \rightarrow j$ by using
equation (13).

$$e_{ij} = d_{ij} - y_{ij} \text{ and } e_{ji} = y_{ij} \qquad (13)$$

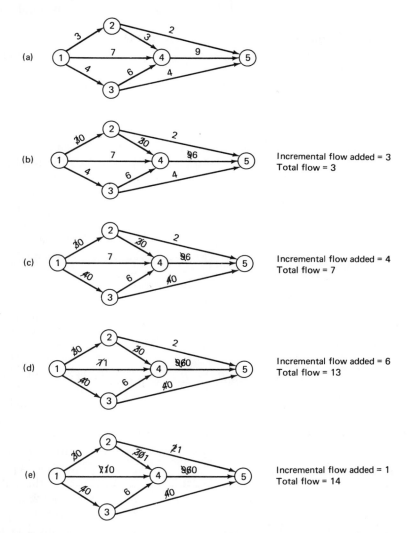

Figure 9-19 Illustrative network with remaining branch capacity indicated.

If one applies this intuitive procedure to the network in Figure 9-20, one could go through the iterations listed in Table 9-5 to arrive at the maximal flow of 14 units; the effects of each of these iterations are indicated in Figure 9-19b, c, d, and e. It should be emphasized that while these particular four iterations are not unique, the end result consisting

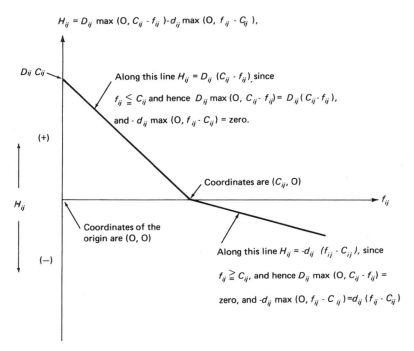

$$H_{ij} = D_{ij} \max (0, C_{ij} - f_{ij}) - d_{ij} \max (0, f_{ij} - C_{ij}),$$

$D_{ij} C_{ij}$

Along this line $H_{ij} = D_{ij} (C_{ij} - f_{ij})$, since

$f_{ij} \leq C_{ij}$ and hence $D_{ij} \max (0, C_{ij} - f_{ij}) = D_{ij} (C_{ij} - f_{ij})$,

and $- d_{ij} \max (0, f_{ij} - C_{ij}) = $ zero.

(+)

Coordinates are $(C_{ij}, 0)$

H_{ij} f_{ij}

Coordinates of the
origin are (O, O)

(−)

Along this line $H_{ij} = -d_{ij} (f_{ij} - C_{ij})$, since

$f_{ij} \geq C_{ij}$, and hence $D_{ij} \max (0, C_{ij} - f_{ij}) = $

zero, and $-d_{ij} \max (0, f_{ij} - C_{ij}) = d_{ij} (f_{ij} - C_{ij})$

Figure 9-20   Development of the Piece-Wise linear convex nature of H_{11}.

of a maximal flow of 14 is unique. A formal procedure for finding this value will now be given.

Table 9-5.   One Possible Set of Iterations in Finding Maximal Network Flow for Figure 9-19

Iteration	Path	$\overline{d}$	Figure Number
1	(1, 2), (2, 4), (4, 5)	3	9-19b
2	(1, 3), (3, 5)	4	9-19c
3	(1, 4), (4, 5)	6	9-19d
4	(1, 4), (4, 2), (2,5)	1	9-19e

Maximal Flow = 14

Node Labeling Algorithm for Finding Maximal Network Flow

Starting with the source, and working towards the sink, the nodes are given a two part label $[k, \epsilon(j)]$ where k indicates the node from which one came to label node j, and $\epsilon(j)$ indicates the minimum, *but positive*, excess branch flow along the path to node j; it is equal to $\epsilon(k)$ or e_{kj}, whichever is smaller. For example, if one is labeling node 5 from node 4, then

$$[k,\epsilon(j)] = \{4, \text{ minimum } [e_{45},\epsilon(4)]\}$$

This labeling process is repeated until one of two cases occur.

(a) No additional nodes can be labeled, and the sink is not labeled. In this case, the existing flow is maximal, and the algorithm is terminated. This case is called nonbreakthrough.

(b) The sink is labeled. In this case, called breakthrough, the existing flow can be increased by an amount equal to $\epsilon(N)$. The flow is increased and we repeat this node labeling algorithm on the network with excess capacities revised as follows:

$$\hat{e}_{rs} = e_{rs} - \epsilon(N), \hat{e}_{sr} = e_{sr} + \epsilon(N), \text{ and}$$
$$\hat{e}_{ij} = e_{ij} \text{ for branches not in the above labeled path}$$

The application of this node labeling algorithm to the network shown in Figure 9-19 is summarized in Table 9-6. The last column in this table

Table 9-6. Node Labels for Network Flow Algorithm Applied to Figure 9-19

Node	9-19b	*Iterations Corresponding to Figures 9-19b, c, d, and e* 9-19c	9-19d	9-19e	—
1	$(-, \infty)$	$(-, \infty)$	$(-, \infty)$	$(-, \infty)$	$(-, \infty)$
2	$(1, 3)$	—	—	$(4, 1)$	—
3	—	$(1, 4)$	—	—	—
4	$(2, 3)$	—	$(1, 7)$	$(1, 1)$	—
5	$(4, 3)$	$(3, 4)$	$(4, 6)$	$(2, 1)$	—

indicates that with excess capacities, as given in Figure 9-19e, no nodes other than the source can be labeled; hence, nonbreakthrough occurs and the algorithm is terminated.

It is clear that flows obtained at each stage in the above labeling algorithm are feasible; it is not obvious, however, that the final flow is maximal. The latter proof, which will only be outlined here, is based on the following definition of a cut set of network branches.

Cut

A cut in a network is a collection of oriented branches such that every oriented path from source to sink contains at least one branch in the cut.

There is an important max flow-min cut theorem which states that if we find the total capacity of the cut, i.e., the sum of the capacities of the branches in the cut set, then the capacity of the minimum capacity cut set is equal to the maximal flow in the network. The proof of this theorem is based on the obvious fact that the total capacity of any cut set is equal to or greater than the total flow in the network, which by definition must pass through the cut set. Now, if one divides the network nodes into two subsets, L denoting labeled nodes, and U the unlabeled nodes, then it can be shown that for the network as labeled at the termination of the above algorithm, the following holds.

$$\sum_{i \in L} \sum_{j \in U} d_{ij} = \text{total capacity of cut set} = \text{total network flow} \quad (14)$$

Here the total capacity of the cut set is equal to the total flow, and hence, it must be the maximal flow, since the network flow can never exceed the total capacity of *any* cut set. The particular set of branches defined by equation (14) is referred to as the minimal cut set, i.e., the cut set having minimal capacity. In the above example, this cut set consists of branches 1-2, 1-4, and 1-3, whose capacity of $3 + 7 + 4 = 14$ is equal to the maximal network flow.

APPLICATION OF NETWORK FLOW THEORY TO THE TIME-COST TRADE-OFF PROBLEM

The linear programming formulation of the time-cost trade-off problem has been given above in equation (4). It was referred to as the primal problem, and was written out in Table 9-4 for the illustrative network given in Figure 9-14. Applying the dual theorem of linear programming to this network leads to the dual formulation given in Table 9-7. In this table the variables corresponding to constraint equations (4b), (4c), and (4d) are denoted by f_{ij}, v_{ij}, and w_{ij}, respectively, while the constraint on the project duration, equation (4e), gives rise to the variable v. Note also that all of the dual constraints are equalities, since all of the primal variables are unrestricted in sign, and the dual variables are all constrained to be nonnegative, since the primal constraints are all inequalities.

The first constraint equation of the dual formulation in Table 9-7 has a *network flow* interpretation, stating that the flow out of the initial (node) event is equal to the total network flow v, and the fourth equation states that the flow into the terminal (node) event is also v. Similarly, the second and third constraint equations state that there is a conservation of flow at the intermediate network events, i.e., at events 2 and 3. It will be shown below that the variables v_{ij} and w_{ij} can be expressed as func-

Table 9-7. Dual Linear Programming Formulation of the Time-Cost Trade-off Problem

Minimize $G = \sum_i \sum_j t_{ij}(0) + \sum_i \sum_j v_{ij}D_{ij} - \sum_i \sum_j w_{ij}d_{ij} + \lambda v$

$\qquad\qquad = \lambda v + \sum_i \sum_j (v_{ij}D_{ij} - w_{ij}d_{ij})$

Subject to

f_{12}	f_{13}	f_{23}	f_{24}	f_{34}	v_{12}	v_{13}	v_{23}	v_{24}	v_{34}	w_{12}	w_{13}	w_{23}	w_{24}	w_{34}	v	= cost coefficient
1	1														−1	= 0
−1		1	1													= 0
	−1	−1		1												= 0
			−1	−1											1	= 0
1					1					−1						= C_{12}
	1					1					−1					= C_{13}
		1					1					−1				= C_{23}
			1					1					−1			= C_{24}
				1					1					−1		= C_{34}

and,

$f_{ij} \geqq 0,\ v_{ij} \geqq 0,\ w_{ij} \geqq 0,\ \text{and}\ v \geqq 0$

tions of the flow variables, f_{ij}, and the activity cost slopes, C_{ij}. Thus, in the last five constraint equations in Table 9-7, the activity cost slopes will be the constraints on the flow in the network activities.

Referring to the dual formulation (since $D_{ij} \geqq d_{ij}$ for all $_{ij}$), v_{ij} and/ or w_{ij} must be 0 in an optimal solution. For example, suppose $(C_{12} - f_{12}) > 0$; then $v_{12} - w_{12} > 0$ must hold because of the constraint which states

$$f_{12} + v_{12} - w_{12} = C_{12}$$

and hence $v_{12} > w_{12}$. Now, since $D_{12} \geqq d_{12}$, minimizing $(v_{12}D_{12} - w_{12}d_{12})$ in the objective function requires that $w_{12} = 0$, since it is constrained to be nonnegative. In this manner, one can verify each of the following statements

$$C_{ij} - f_{ij} > 0 \rightarrow w_{ij} = 0 \tag{15}$$

$$C_{ij} - f_{ij} = 0 \rightarrow v_{ij} = w_{ij} = 0 \tag{16}$$

$$C_{ij} - f_{ij} < 0 \rightarrow v_{ij} = 0 \tag{17}$$

Hence, one may eliminate the v_{ij} and w_{ij} variables by expressing them as functions of the known cost slopes, C_{ij}, and the remaining dual variable, f_{ij}.

From equations (15), (16), and (17), one can express v_{ij} and w_{ij} in terms of C_{ij} and f_{ij} as follows:

$$v_{ij} = \text{maximum } (0, C_{ij} - f_{ij}), \text{ and} \tag{18}$$

$$w_{ij} = \text{maximum } (0, f_{ij} - C_{ij}) \tag{19}$$

The objective function of the dual then becomes

$$\text{Minimize } G = \lambda v + \sum_i \sum_j [D_{ij} \max(0, C_{ij} - f_{ij}) - d_{ij} \max(0, f_{ij} - C_{ij})] \tag{20}$$

While the v_{ij} and w_{ij} variables have been eliminated, the resulting problem is now a nonlinear programming problem because of the nonlinearity of the objective function, G. Fortunately, a function of the form

$$H_{ij} = D_{ij} \max(0, C_{ij} - f_{ij}) - d_{ij} \max(0, f_{ij} - C_{ij}) \tag{21}$$

is piece-wise linear and convex since $d_{ij} \leqq D_{ij}$. This is shown in Figure 9-20. Thus, even though equation (20) above is nonlinear, it can be dealt with by linear methods merely by breaking each of the variables, f_{ij}, into two parts, corresponding to the two pieces of their cost curve. This latter procedure is not unlike our previous replacement of a convex activity cost curve by two or more pseudo-activity cost curves. Here, one can replace f_{ij} by the sum of two nonnegative variables as follows:

$$f_{ij} = f_{ij1} + f_{ij2} \tag{22}$$

where the new variables are subject to the constraints

$$0 \leq f_{ij1} \leq C_{ij}, \text{ and } 0 \leq f_{ij2} \leq \infty \tag{23}$$

To verify that this formulation is correct, note first that the coefficient of f_{ij1} in the H_{ij} portion of the dual objective function is $- D_{ij}$, and for f_{ij2} it is $- d_{ij}$. Since the coefficient of f_{ij2} is greater (algebraically) than the coefficient for f_{ij1}, it follows that in minimizing the dual objective function, if $f_{ij2} > 0$, then $f_{ij1} = C_{ij}$, its maximum value, and hence the replacement is physically valid, i.e., until f_{ij1} reaches C_{ij}, its maximum value, $f_{ij2} = 0$, and when $f_{ij2} > 0$, $f_{ij1} = C_{ij}$, a constant. Now if one defines C_{ijk} and d_{ijk} as suggested by the above analysis

$$\begin{array}{llll} C_{ij1} &= C_{ij} & \text{and} & C_{ij2} = \infty \\ d_{ij1} &= D_{ij} & \text{and} & d_{ij2} = d_{ij} \end{array} \tag{24}$$

then one can rewrite the dual formulation once again in terms of the illustrative example as

$$\text{Minimize } G = \lambda v - \sum_i \sum_j \sum_k f_{ijk} d_{ijk} \tag{25a}$$

$$\text{Subject to } \sum_j \sum_k (f_{ijk} - f_{jik}) = \left\{ \begin{array}{l} v; i = 1 \\ 0; i = 2, 3 \\ - v; i = 4 \end{array} \right. \tag{25b}$$

$$\begin{array}{l} 0 \leq f_{ijk} \leq C_{ijk}; \text{ all ijk} \\ 0 \leq v \end{array} \tag{25c}$$

This formulation has the following *network flow* interpretation. First, enlarge the project by paralleling each original activity by a second activity labeled *ij*1 and *ij*2, respectively, as shown in Figure 9-21. Each activity flow, f_{ijk}, has a maximum capacity of C_{ijk}, and an objective function coefficient of d_{ijk}, i.e., the activity normal or crash time, and the total flow, v, has an objective function coefficient of λ, the project duration time. *The problem now is to construct a total flow of value, v, from the initial to the terminal events in the new network, that minimizes G, the objective function.*

Except for the details that minimization has replaced maximization and pairs of activities join events, the problem is now a standard linear programming problem. The first difference could be eliminated by multi-

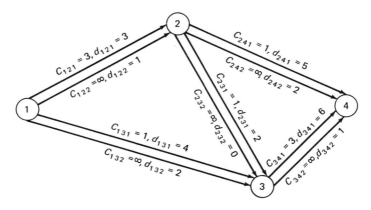

Figure 9-21 Enlarged project network.

plying the objective function by $- 1$, and the latter difference by replacing each pair of parallel activities by two pseudo-activities in series. These changes need not be made here because the problem is now in the desired form for developing the *network flow* algorithm treated below. Before proceeding with this development, however, the nature of the simplex solution to this problem will be considered.

Simplex Solution Procedure

If one were to solve this problem using the simplex method of linear programming, one could multiply the objective function by $- 1$ to change it to a maximization problem as follows:

$$- G = G^* = - \lambda v + \sum_i \sum_j \sum_k f_{ijk} d_{ijk} \qquad (26)$$

and begin the computations by letting $\lambda = \lambda_D$, the length of the project critical path using all normal times, D_{ij}. Next, one could treat the problem as a parametric problem in linear programming. Here one finds out how much λ can be changed without changing the basic feasible solution, and subsequently goes from one basic solution to another until $\lambda = \lambda_d$, the minimum possible length of the project critical path(s), i.e., the project duration when all critical path activities are scheduled at their crash times, d_{ij}.

To see what is going on here, consider the problem with only two variables, v and f, and objective function, $G^* = - \lambda v + df$. Suppose

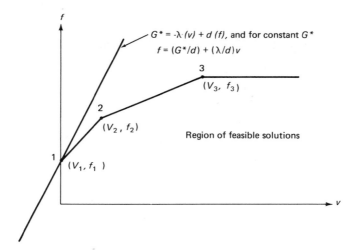

Figure 9-22 Simplified version of time-cost trade-off problems to illustrate parametric linear programming.

further that the constraints on the problem are as shown in Figure 9-22. Heuristically, it can be seen from this figure that for a fixed value of λ, which makes G^* parallel to the line 1-2, the flow, v, can be increased from v_1 to v_2. When point 2 has been reached, however, λ must be reduced in order for G^* to be parallel to the line 2-3. At this point, v can again be increased, this time from v_2 to v_3. The network flow algorithm to be described below will go through a series of such steps. Now, suppose the current solution is at the point (v_1, f_1). As λ is decreased in the range to go from point 1 to 2, the value of the objective function increases linearly at rate $v = v_1$, and when the basic feasible solution changes from (v_1, f_1) to (v_2, f_2), the rate of change of the objective function increases for further reductions in λ because $v_2 > v_1$. The effect of this is to produce a total project direct cost curve composed of a series of straight lines of constantly increasing slope as λ decreases. This type of curve is shown in Figure 9-23 for the solution to the illustrative network given in Figure 9-14.

Using the simplex algorithm to solve this problem would severely limit the size of the project network that could be handled, since the number of variables in the problem is twice the number of activities in the network and the number of constraints is even greater. Fortunately, the problem can be formulated as a *network flow* problem and solved using the flow algorithm described below.

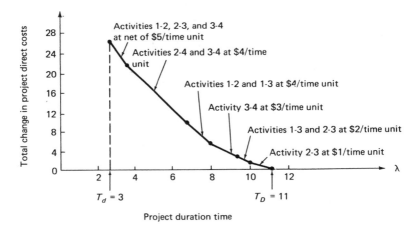

Figure 9-23   Total project direct time-cost curve for illustrative example.

THE DUAL NETWORK FLOW ALGORITHM

This algorithm is an event labeling process similar to the one developed in the previous section on Network Flow Theory. The algorithm is essentially a systematic search for a flow augmenting path from the source event (node) denoted by s, to the terminal (sink) network event (node), denoted by t. An event is considered to be in one of three states: (1) unlabeled, (2) labeled and scanned, or (3) labeled and unscanned. One main reason underlying the computational efficiency of the labeling process is that once an event is labeled and scanned, it can be ignored for the remainder of that labeling cycle. Labeling a node i corresponds to locating a path from event s to event i that can be the initial segment of a flow augmentation path. While there may be many such paths from s to i, finding one suffices. Enough information is carried along in the labels so that if the terminal event, t, is labeled, a case called breakthrough, the resulting flow change along the path can be made readily. If, on the other hand, the terminal event is not labeled, a case called nonbreakthrough, the flow through the network is maximal, and the set of activities leading from labeled to unlabeled activities, i.e., the minimal cut-set, determines what event times must be reduced and by how much they may be reduced before the next labeling cycle is undertaken. This corresponds to moving from point (v_1, f_1) to (v_2, f_2) in Figure 9-22. These event time reductions, which always include the terminal event, determine the total project duration reduction, as well as the reduction in the individual project activities, and hence the increase in the project direct costs can be computed. This point is expressed analytically in equation (28) below.

First, it is assumed that the starting project event is numbered $s = 1$, and the terminal event is numbered t, and the intermediate events are numbered so that all activities ij, $i < j$. Now, one may let

$$\overline{d}_{ijk} = T_i + d_{ijk} - T_j \qquad (27)$$

Then the activities for which $\overline{d}_{ijk} = 0$ are said to be *admissible*. Note, that $\overline{d}_{ij1}$ denotes the negative of the total slack for activity ij when the latter is scheduled at its normal time ($d_{ij1} = D_{ij}$), and $\overline{d}_{ij2}$ has a similar interpretation when the activity is scheduled at its crash time ($d_{ij2} = d_{ij}$).

Start

(a) Let $\lambda = \lambda_D = T_t$, computed in the usual manner letting $y_{ij} = D_{ij}$ for all ij.

(b) Let all $f_{ijk} = 0$.

At this point, one is ready to begin the first of two labeling steps. A label, assigned to node j, is of the form $[i, \ k^{+ \text{ or } -}, \ \epsilon(j)]$. The first element, i, denotes the node from which the labeling is being done. The second element, which is 1^+, 1^-, 2^+, or 2^-, denotes the path and direction of the labeling; $k = 1$ or 2 denotes the path and the plus sign denotes that the labeling (flow) is from i to j, whereas the minus sign denotes that the labeling (flow) is from j to i. The last element, $\epsilon(j)$, denotes the largest permissible flow change along the path from s to j.

First Labeling

Always label the initial project event $s = 1$ with $[-, -, \epsilon(1) = \infty]$. Event j can be labeled from a labeled node i if activity $(i, j, 2)$ is admissible; event j is then labeled $[i, 2 +, \epsilon(j) = \infty]$.

If breakthrough occurs (event t gets labeled), then terminate; otherwise, go to the second labeling†

Second Labeling

The above event labels are retained, and all events revert to the unscanned state. When scanning a labeled event i, the labeling rules are, j can be labeled if (a) or (b) is satisfied.

(a) ijk is admissible and $f_{ijk} < C_{ijk}$. Here event j is labeled $[i, k^+, \epsilon(j)]$ where $\epsilon(j) = \min [\epsilon(i), C_{ijk} - f_{ijk}]$.

† This termination rule follows from the fact that breakthrough here signifies that a critical path through the network has been found with all of its activities at their crash times; $\overline{d}_{ijz} = 0 \rightarrow y_{ij} = d_{ij}$ for activity ij. Hence, no further reduction in the project duration time is possible.

(b) *jik* is admissible and $f_{jik} > 0$. Here event j is labeled $[i, k-, \epsilon(j)]$
where $\epsilon(j) = \min [\epsilon(i), f_{jik}]$. (Note, $j < i$ in this instance.)

If breakthrough occurs at the completion of the second labeling, change the flow by adding or subtracting $\epsilon(t)$ (according to the sign superscript on k in the event label) along the path from s to t directed by the labels. If nonbreakthrough occurs, single out the following subsets of activities which together comprise the minimal cut-set for the network.

$S_1 = [(ijk) \mid i$ labeled, j unlabeled, $\overline{d}_{ijk} < 0]$ and let $\Delta_1 = \min_{S_1} [- \overline{d}_{ijk}]$

$S_2 = [(ijk) \mid i$ unlabeled, j labeled, $\overline{d}_{ijk} > 0]$ and let $\Delta_2 = \min_{S_2} [\overline{d}_{ijk}]$

and let $\Delta = \min (\Delta_1, \Delta_2)$. Now change event times, T_i, by subtracting Δ from all T_i corresponding to unlabeled i. Discard all of the old event labels and start a new cycle of this labeling process until it is terminated by a breakthrough in a First Labeling.

Each new set of event times, T_i, yield a new point on the project time-cost curve. Letting $P(\lambda)$ denote the total project direct costs for a scheduled project duration of $\lambda = T_t$, one can write

$$P(\lambda) = \sum_i \sum_j K_{ij} - \sum_i \sum_j C_{ij}y_{ij} = K - \sum_i \sum_j C_{ij}y_{ij} \quad (28)$$

where $y_{ij} = \min [D_{ij} (T_i - T_i]$

Application of the Dual Network Flow Algorithm

The above network flow algorithm will now be applied to the simple network given in Figure 9-14 at the beginning of this appendix, and shown in enlarged form in Figure 9-21. The starting conditions for the labeling are listed in Table 9-8 under the initial labeling cycle column, i.e., $f_{ijk} = 0$ for all ijk, and $\overline{d}_{ij1}$ denotes the negative of the total slack for activity ij as determined in the initial forward pass computations, and $\overline{d}_{ij2}$ is merely $\overline{d}_{ij1} - (D_{ij} - d_{ij})$. The initial cycle of Second Labeling results in a breakthrough, and hence the flow is augmented by $\Delta = 1$ unit along the path indicated by the node labels, i.e., activities 3, 4, 1, then 2, 3, 1, and finally 1, 2, 1; this augmentation is written in the labeling cycle I column. In labeling the nodes on labeling cycle I, the Second Labeling results in nonbreakthrough and calls for a reduction in the unlabeled event times, i.e., events 3 and 4. The updated event times, scheduled activity durations, and activity direct cost changes are given in the tabulation below. This process is continued until the First Labeling of cycle XI results in a breakthrough, which signals the termination of the algorithm. This is equivalent to saying that the activities along one or more legs of the critical paths have all reached their crash times, and hence no further reduction in the project duration time is possible. The project cost curve developed in making these computations is given in Figure 9-23.

Labeling Cycle	Initial	I	II	III	IV	V	VI	VII	VIII	IX	X	END
T_1	0	0	—	0	—	0	—	0	0	—	0	—
T_2	3	3	—	3	—	3	—	2	2	—	1	—
T_3	5	4	—	3	—	3	—	2	2	—	2	—
$T_4 = \lambda$	11	10	—	9	—	8	—	7	4	—	3	—
y_{12}	3	3	—	3	—	3	—	2	2	—	1	—
y_{13}	4	4	—	3	—	3	—	2	2	—	2	—
y_{23}	2	1	—	0	—	0	—	0	0	—	1	—
y_{24}	5	5	—	5	—	5	—	5	2	—	2	—
y_{34}	6	6	—	6	—	5	—	5	2	—	1	—
f_{12}	0	1	—	1	—	2	—	3	3	—	3	∞
f_{13}	0	0	—	1	—	1	—	1	1	—	2	∞
f_{23}	0	1	—	1	—	2	—	2	2	—	1	1
f_{24}	0	0	—	0	—	0	—	1	1	—	2	∞
f_{34}	0	1	—	2	—	3	—	3	3	—	3	∞
v	0	1	—	2	—	3	—	4	4	—	5	—
Δ Cost †	0	1	—	2	—	3	—	4	12	—	5	—
Cumulative Δ Cost	0	1	—	3	—	6	—	10	22	—	27	—

$$\dagger\,\Delta \text{ Cost} = \left(\sum_i \sum_j y_{ij}C_{ij} \right)^{\text{previous}}_{\text{cycle}} - \left(\sum_i \sum_j y_{ij}C_{ij} \right)^{\text{current}}_{\text{cycle}}$$

Table 9-8. Event Labeling

Node	$i, k, \epsilon(j)$	$i, k, \epsilon(j)$	$i, k, \epsilon(j)$	$i, k, \epsilon(j)$	$i, k, \epsilon(j)$
$1 = s$	$-, -, \infty$	$-, -, \infty$	$-, -, \infty$	$-, -, \infty$	$-, -, \infty$
2	$1, 1^+, 3$	$1, 1^+, 2$	$1, 1^+, 2$	$1, 1^+, 2$	$1, 1^+, 2$
3	$2, 1^+, 1$	—	$1, 1^+, 1$	—	$2, 2^+, 2$
$4 = t$	$3, 1^+, 1$	—	$3, 1^+, 1$	—	$3, 1^+, 1$

Labeling Cycle	Initial	I	II	III	IV
Activity 1–2					
f_{121}	0	1			
f_{122}	0				
$C_{12} = 3 \quad \bar{d}_{121}$	0				
$\bar{d}_{122}$	−2				
Activity 1–3					
f_{131}	0			1	
f_{132}	0				
$C_{13} = 1 \quad \bar{d}_{131}$	−1		0		1
$\bar{d}_{132}$	−3		−2		−1
Activity 2–3					
f_{231}	0	1			
f_{232}	0				
$C_{23} = 1 \quad \bar{d}_{231}$	0		1		2
$\bar{d}_{232}$	−2		−1		0
Activity 2–4					
f_{241}	0				
f_{242}	0				
$C_{24} = 1 \quad \bar{d}_{241}$	−3		−2		−1
$\bar{d}_{242}$	−6		−5		−4
Activity 3–4					
f_{341}	0	1		2	
f_{342}	0				
$C_{34} = 3 \quad \bar{d}_{341}$	0				
$\bar{d}_{342}$	−5				

Using Network Flow Algorithm

$i, k, \epsilon(i)$	$i, k, \epsilon(i)$	$i, k, \epsilon(i)$	$i, k, \epsilon(i)$	$i, k, \epsilon(i)$	$i, k, \epsilon(i)$	$i, k, \epsilon(i)$
$-, -, \infty$	$-, -, \infty$	$-, -, \infty$	$-, -, \infty$	$-, -, \infty$	$-, -, \infty$	$-, -, \infty$
$1, 1^+, 1$	$1, 1^+, 1$	—	$3, 2^-, 1$	$3, 2^-, 1$	—	$1, 2^+, \infty$
$2, 2^+, 1$	$2, 2^+, 1$	—	$1, 2^+, \infty$	$1, 2^+, \infty$	$1, 2^+, \infty$	$1, 2^+, \infty$
—	$2, 1^+, 1$	—	—	$2, 2^+, 1$	—	$2, 2^+, \infty$

V	VI	VII	VIII	IX	X	END
2		3				
			1			2
			−1			0
					1	
			2			
			0			
1					0	
						1
						−1
		1			1	
	0			3		
	−3			0		
3						
	1			4		5
	−4			−1		0

10

NETWORK COST CONTROL

In the previous chapter, techniques are explained for planning a project to optimize its time-cost restraints. This chapter considers means of controlling the expenditures as the project progresses in time and in accomplishment. The use of the network and critical path schedule computations as vehicles for expenditure status reports is discussed, and analysis of these reports to aid in the control of future expenditures is illustrated. Such status reporting and analysis may take many forms; this chapter shall consider primarily the following basic questions:

(1) What are the actual project costs to date?
(2) How do the actual costs to date compare with planned costs to date?
(3) What are the project accomplishments to date?

(4) How do the actual costs of specific accomplishments compare with the planned costs of these accomplishments?

(5) By how much may the project be expected to overrun or under-run the total planned cost?

(6) How do the above questions apply to various subdivisions and levels of interest within the project?

Systems that answer these questions employ what is sometimes referred to as an "enumerative cost model," as distinguished from the optimization cost models treated in Chapter 9. Some of the problems that arise in developing enumerative systems are discussed in this chapter, along with suggestions and illustrations of several existing systems.

BACKGROUND

The network diagram with costs associated with each activity or group of activities was recognized early as a potential means of improving cost control as well as schedule control of projects. However, one of the curious aspects of the development of network systems is that the inclusion of cost control features has lagged far behind other supplements to the basic scheduling features. While the more technically complex areas of time-cost trade-off and statistical schedule predictions were developed along with the earliest CPM and PERT systems, the relatively simple and more generally applicable features related to routine management of project expenditures have only recently received widespread attention. A few computer programs and manual procedures to answer the above questions were developed by individual CPM and PERT users in the period 1959–1962. In 1962 a major boost to the interest in network cost control was provided by agencies of the U. S. Government. The Department of Defense and the National Aeronautics and Space Administration jointly issued a manual entitled *DOD and NASA Guide, PERT Cost Systems Design*,[4] which emphasized the cost control aspects of "PERT-type systems." Several companies and agencies in the aerospace field had already been working with various PERT Cost procedures and computer programs, but the *DOD and NASA Guide* served to formalize the interest of the government and thus to initiate active development of the procedures throughout the aerospace industry. My mid-1963 the use of PERT Cost procedures had become a requirement in certain military research and development projects, and several new computer programs were being written.

In the construction industry interest in the cost control potential of CPM has grown more slowly, although there is no essential difference in CPM and PERT with respect to cost control. In 1963 the U.S. Army Corps of Engineers provided some impetus in this direction by formalizing its interest in "network analysis systems" for schedule and cost control.[5] A few state highway departments, universities, architect-engineers, and other organizations administering construction contracts have also begun to specify some applications of cost data along with CPM requirements. Some construction companies have developed CPM cost control systems of their own.[3]

BASIC PROBLEMS

While in theory the concepts of cost control based on the project network are not complex, the design and the implementation of a practical cost control system is not readily accomplished. The fundamental problems facing the system designer may be classified as (1) those related to organizational conflicts, and (2) those related to the necessary efficiency of the system. The basic organizational problem is the conflict between the project approach of network cost control and the functional approach of cost accounting procedures found most in industry. This conflict is manifest particularly in the design of input and output phases of network control systems. The input to a network system requires the development of an *activity accounting* procedure by which actual expenditure data are coded to provide association with activities (or groups of activities) in the project network. The output from the system likewise must be project-oriented to provide *project summary reports,* organized by time period, areas of responsibility, and technical subdivisions of the project.

The efficiency of the system is a problem because network-oriented systems lend themselves to major increases in the amount of detail available to the manager. The level of detail is both the promise and the inherent hazard of such systems, and it is one of the primary tasks of the system designer to achieve the level of detail that provides the greatest return on the investment in the system. A network cost control system can easily require routine input data in quantities and frequencies that project personnel find extremely burdensome. Unless the requirements are reduced and the procedures simplified, the system will come to an early death. Similar dangers lie in the design of the data-processing and output phases of the system.

ACTIVITY ACCOUNTING

Basic CPM and PERT systems gained acceptance rapidly because the critical path concept filled a generally recognized need for improved, formal procedures for project planning and scheduling. In the cost control area, however, formal accounting procedures have been long established. Thus, one of the greatest obstacles to the use of network cost control, particularly in the large aerospace firms, has been the difficulty of developing a network system compatible with the established accounting system.

Generally speaking, accounting systems in organizations engaged in project-type endeavors are designed to plan budgets and to report expenditures both by organizational unit and by project. However, the emphasis is on the organizational unit accounts (section, department, division, etc.), inasmuch as the objectives of the accounting system are summaries of expenditures by the functional elements of the organization. Where the accounting codes also permit summaries by project, the project summary represents the lowest level of detailed cost reporting available to the operating management. The purpose of the critical path approach, on the other hand, is to provide more detailed information and control *within* the project.

Cost data, like time data, must be applied to *activities* in the project network. The budgeting and recording of expenses both by cost item and by network activity clearly means that a more elaborate system is required. Not only are there more codes and figures to deal with, but there are many more decisions to make. For example, consider the following typical questions developed in the application of cost data to a project network:

(1) Electronic testing gear is purchased for use in several activities in the project. Should the cost of the gear be assigned entirely to the purchasing activity, when the expenditure actually occurs, or should it be allocated over the activities involving use of the gear?

(2) What cost, if any, should be assigned to the curing of concrete? Approval of shop drawings? Negotiation of a subcontract?

(3) Should overhead be included or only direct costs? If overhead is included, is it computed the same way for all activities?

(4) How should the costs associated with project management be shown, as activities or as overhead?

Such questions arise largely from the fact that basic time-oriented networks and the list of project cost items represent two different sets of data. Although the sets largely overlap, many elements of the project which involve costs have not been shown in the network. This is particularly true of management and other overhead expenses. Certain other activities involve no direct costs but consume time and perhaps should account for a portion of the indirect costs. Various answers to these activity accounting problems have been worked out in the development of the network cost control systems that exist today. One of the most significant of the current techniques is the use of groups of activities, or "work packages," in the coding of the cost accounts. Work packages are selected groups of activities that define a particular unit of work under the responsibility of a particular organizational unit. Cost data for work packages, including supervisory and management overhead, are summarized by the package code. The *DOD-NASA Guide*† includes the following comments on the organization of work packages:

"*End Item Subdivisions.* The development of the work breakdown structure begins at the highest level of the program with the identification of project and items (hardware, services, equipment, or facilities). The major end items are then divided into their component parts (e.g., systems, subsystems, components), and the component parts are further divided and subdivided into more detailed units. . . . The subdivision of the work breakdown continues to successively lower levels, reducing the dollar value and complexity of the units at each level, until it reaches the level where the end item subdivisions finally become manageable units for planning and control purposes. The end item subdivisions appearing at this last level in the work breakdown structure are then divided into major work packages (e.g., engineering, manufacturing, testing). At this point, also, responsibility for the work packages will be assigned to corresponding operating units in the contractor's organization.

"The configuration and content of the work breakdown structure and the specific work packages to be identified will vary from project to project and will depend on several considerations: the size and complexity of the project, the structure of the organizations concerned, and the manager's judgment concerning the way he wishes to assign responsibility for the work. These considerations will also determine the number of end item subdivisions that will be created on the work breakdown structure before the major work packages are identified and responsibility is assigned to operating units in a contractor's organization.

"*Further Functional or Organizational Subdivisions.* An organization unit will usually identify smaller work packages within the major work packages assigned to it. This division of work may take the form of more detailed functional (e.g., engineering) identification, such as systems engineering, electrical engineering, mechanical engineering, or it may take the form of a more detailed end item identification with engineering, such as instrumentation engineering, power cable engineering, missile section assembly engineering, and so forth. . . . The form chosen for more detailed identification will depend, again, on the structure of the performing organization and the manager's judgment as to the way he wishes to assign responsibility for the work.

† Reference 4, pages 27 and 29.

The *number* of these smaller subdivisions will naturally depend on the dollar value of the major work packages and the amount of detail needed by the manager to plan and control his work. Normally, the *lowest level* work packages will represent a value of no more than $100,000 in cost and no more than three months in elapsed time.

"THE WORK PACKAGES FORMED AT THE LOWEST LEVEL OF BREAKDOWN, THEN, CONSTITUTE THE BASIC UNITS IN THE PERT COST SYSTEM BY WHICH ACTUAL COSTS ARE (1) COLLECTED AND (2) COMPARED WITH ESTIMATES FOR PURPOSES OF COST CONTROL."

The work package approach represents a practical compromise in the concept of activity accounting in the aerospace industry. A similar approach may be applicable in other industries where network cost control is desired. In the construction industry, for example, the lowest level work packages may be identical or nearly identical to the bid items in the project. In some cases the package would be represented on the network by a single activity; in most cases it would be comprised of a group of activities within a division of the project, such as the concrete work for the first floor of a building. The specific work package organization would naturally vary with the type of project, areas of responsibility of the foremen and superintendents, the capabilities of the cost accounting system, and other factors related to the particular company and its projects. An explanation of one construction company's approach is given at the end of this chapter.

DEVELOPING THE COST PLAN, OR BUDGET

Assuming that an activity accounting scheme has been established, the first step in the application of network cost controls to a project is the preparation of the network. While the network rules are not changed by cost considerations, the organization of the network will be influenced by the work package approach. The next step is the basic scheduling computation which gives the earliest and latest start and finish times for each activity.

At this point the estimated cost data for each activity or work package may be added to the network, and the first cost computation is made, which is the summation of all estimated costs by time period. This summation is performed in the same way that "resources" were summed in Chapter 8. In this summation it may be assumed that all activities will begin at their earliest start times, ES. One may also elect to make the summation based on latest start times, LS. Using Figure 10-1 as an example, these computations result in the curves shown in Figure 10-2, which define the range of feasible budgets.

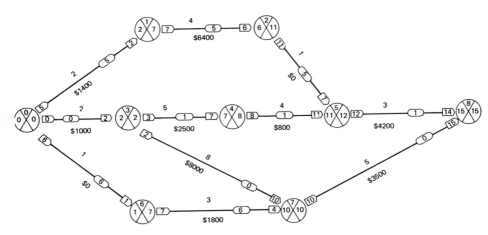

Figure 10-1 Illustrative network with estimated costs of activities (or work packages). Time unit is one week.

Now a particular budget curve, lying somewhere between the *ES* and *LS* curves, is usually desired. Such a curve may be determined by inspecting the network for activities which, for one reason or another, should start at some time between their *ES* and *LS* times. The budget curve then may be recomputed on the basis of the scheduled start times.

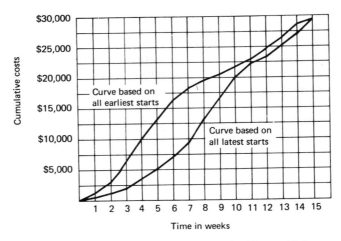

Figure 10-2 Cumulative cost curves for network in Figure 10-1.

In some cases it may be desirable to select the scheduled start times by systematic procedures, such as those described in Chapter 8. Such an approach may be warranted by special problems in scheduling labor or equipment. The same procedures explained in Chapter 8 may also be applied, where the rate of expenditure of funds is the critical limitation. For example, it may be necessary to find a schedule which limits the maximum expenditure in any one time period; in this case extensions in the time length of the critical path may be required. In another situation it may be desirable to schedule the project to level the requirements for funds within a limitation on the length of the project. The procedures for both approaches, as given in Chapter 8, are applicable when the limiting resource is money, where money can be considered merely as another type of resource whose availability by time periods is restricted.

OBTAINING ACTUAL TIME AND COST DATA

As the project progresses, actual expenditures are recorded by activity. In the case of labor costs, these may be recorded on weekly time cards which require the distribution of each person's time over the various activities in which he participated. The activities must be coded, of course, to correspond to the network codes. This requirement adds to the paperwork burden of the working personnel and their supervisors, a fact which must be carefully considered in the design of the forms and procedures.

In this regard it should be noted that it is quite possible and desirable to utilize the same input data for payroll and other accounting purposes, thus minimizing the red tape for the personnel by avoiding input duplication. This also implies the development of a mechanized cost accounting system for the organization, or the integration of an existing mechanized system with the network cost control system. An integrated, mechanized cost system of this type has the additional advantage of providing quick reports on actual expenditures. One cause of failure in attempts to develop a network cost control system has been the inability of the existing accounting system to provide actual costs soon enough to be of value in controlling the project.

In the case of materials, subcontracted activities, and other cost items, actual expenditures may be recorded in relation to the total progress on the activity. For example, at the time for the periodic cost report a particular activity may be judged to be 75 percent complete. If this is a subcontracted activity, the cost may be recorded at 75 percent of the total budgeted or contracted cost for the activity. This method involves an assumption of a linear time-cost relationship which may not be appropriate in all cases. In some activities, for example, 75 percent of the costs may be incurred in the first 25 percent of the time required for the activity; the activity of pouring and curing concrete would be an example. In such cases it may be desirable to divide the activity into two or more activities which may each be treated linearly, or some method other than linear apportionment may be used. Naturally, this point may be of considerable concern if the cost control system is to be the basis for periodic payments to subcontractors, or if the financing agent or the contracting agency uses the system as a basis for periodic payments to the prime contractor.

Essentially, the problem here revolves about the difference in the two ways of measuring progress—by time and by costs. It is quite possible that a system will be developed that utilizes both measures. For each activity in progress at the reporting date, estimates of the time and cost remaining will be made. Thus, where a percentage of the time for an activity remains and some other percentage of the costs remain, these data may be recorded in the input forms, and computational use may be made of them. However, it does not appear that complications of this type will be worthwhile. In most applications of network cost controls, at any reporting date only a small percentage of the project activities will be in progress. Among these few activities, most will be adequately treated by the assumption of a linear relationship between the time and cost. Where this assumption is incorrect, the effect on the total cost control pictures is likely to be quite small. By this reasoning then, it is adequate to estimate only the time remaining for an activity in progress. The cost remaining can be taken simply as the difference between the total estimated for the activity and the actual to date.

The input system should, however, permit revisions in the cost estimates for any activity not yet completed. Cost revisions should be treated just as time revisions in the periodic up-dating computations; that is, the new estimates become the basis for the schedule computations and for comparisons with actual time and funds expended.

COST COMPUTATIONS AND PROJECTIONS

Upon receipt of all input data on the progress and expenditures as of the reporting date, and any revised estimates of durations or costs of activities not completed, computations may be made to answer the questions posed at the beginning of this chapter. These computations are the following:

(1) Summation of all actual costs.

(2) Summation of budgeted costs at this point in time.

(3) Summation of budgeted costs for all activities completed and partial costs of activities partially completed. This figure is called "planned cost of work completed," or "value of work."

(4) Computation of difference in actual costs and planned cost of work completed; computation of both dollar amount and percentage of planned cost of work completed.

(5) Computation of a projection of ultimate cost of the total project.

Note that no comparison of costs (1) and (2) is suggested; the comparison indicated in (4) is considered more significant. Of course, any comparisons could be computed, but it is undesirable to clutter the output report with rarely used data, which could easily be computed mentally when needed.

The last step, projecting the ultimate costs of the project, is generally performed by summing the actual costs to date and the latest revised cost estimates of work not yet completed or started. This procedure requires that cost estimates be reviewed by management and revised as may be appropriate, prior to the computation of the new projection. The sum of actuals to date and latest revised estimates of future costs may then be compared with the contracted total cost for the project, to indicate the amount of the expected overrun or underrun.

Another approach, which is not used by any system known to the authors, would involve analysis of the trend in costs by means of some arbitrary formula. For example, one could compute the percentage (over) underrun to date, and weight this figure by the fraction of time consumed along the critical path. To illustrate, suppose a computation is made after 25 percent of the time along the critical path had been consumed, and the actual costs total 20 percent more than the planned costs of the work completed. Using the weighted percentage rule described above, the predicted final cost would be $(.25)(.20) = .05$, or 5 percent over the estimated total cost. Suppose after 90 percent of the critical path is completed, the costs are still running 20 percent above the budget. The prediction now would be $(.90)(.20) = .18$, or an 18 percent overrun. The assumptions underlying this scheme are that minor overruns and underruns will tend to cancel each other in the long run and that a difference early in the project is less significant than one late in the project.

Such an analysis may have practical advantages where the project is composed of a number of repetitive cycles, such as the construction of a bridge, a high-rise building, or pilot production of a missile. In these cases the cost of each cycle is a random variable, and the control of total project costs depends on monitoring the trend of variation in costs of each cycle. There is also some advantage in using a computer to make these computations, for there is a certain psychological effect inherent in a computer prediction. In the main, however, the procedure based on revised estimates of cost to completion is less arbitrary and should provide the most accurate and effective cost control reports.

EXAMPLE

The basic cost control computations can be illustrated by taking the project network in Figure 10-1 and assuming that the management wishes to start each activity at its earliest start time. Hence, the earliest-start curve in Figure 10-2 represents the initial plan for expenditures over the duration of the project. Now suppose that at the end of the fourth week an assessment is made of the progress to date, in terms of both the schedule and the planned expenditures. The status of the project may be recorded in a form similar to Table 10-1, and illustrated

Table 10-1.   Progress Data for Illustrative Project After the Fourth Week (Input to the System)

| | | COMPLETED ACTIVITIES | | |
Activity No.	Date Started	Date Finished	Actual Duration	Actual Cost (dollars)
0–1	0	1	1	1600
1–2	1	4	3	6600
0–3	0	2	2	800
0–6	2	4	2	0
		ACTIVITIES IN PROGRESS		
Activity No.	Date Started	Duration to Date	Estimated Time To Complete	Cost to Date
2–5	4	0	1	0
3–4	2	2	4	1200
3–7	2	2	5	1800
6–7	4	0	3	0

graphically on the network as shown in Figure 10-3. (The reader will note that the addition of actual time and cost data begins to crowd the network presentation, even though some information in Table 10-1 is omitted from Figure 10-3. If the node method is used, the node symbols provide a more efficient means of recording these data. The user

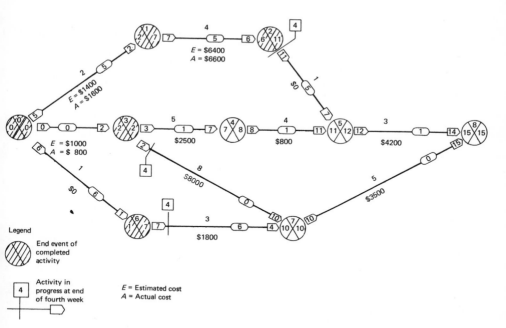

Figure 10-3 Progress at the end of week four, as denoted graphically on the project network.

may also draw the networks larger to provide more space for notation, develop his own notational shorthand, and/or depend primarily on records separate from the network to retain historical information.)

The information in Table 10-1, when compared with the initial network, reveals the following facts:

(1) Activities 0-1 and 1-2 were finished more quickly but also at greater expense than estimated. Since this path has five weeks of slack, it appears that unnecessary expenditures may have been incurred to accelerate these activities.

(2) Activity 0-3 was completed on schedule at less expense than estimated.

(3) Activity 0-6 was delayed in starting, but this had no effect on the schedule or the costs.

(4) Activity 3-7, which started on time and is now in progress, is estimated to require a total of 7 weeks instead of 8 weeks, and the rate of expenditure is less than initially estimated—approximately $900 per week instead of $1000 per week. The original critical path has thus been shortened by one week.

(5) Activity 3-4, which started on time and is now in progress, is taking longer and costing more than estimated. It is now on the critical path.

It is especially noteworthy that the above facts are the result of inspection of the input data. No computer program or even a manual computational procedure, other than mental arithmetic, was required. The important features of the analysis are that *a plan was developed in the beginning and that sufficient records were kept to compare the plan with actual progress by activities.* When the proper data are provided, the analysis is relatively simple.

Now it may also be desirable to provide certain summarizations of the status and predicted ultimate costs of the project. One format of a summary cost control report is shown in Table 10-2. In this table, note the difference between the planned cost to date and the planned cost of work completed. The planned cost to date refers to the *ES* curve in Figure 10-2 which was taken as the plan; at the end of the fourth week $10,000 was to have been spent. Due to certain activities having been completed more quickly than estimated, and at higher costs, the total cost is $2000 greater than expected at the end of the fourth week. However, with regard to the activity cost estimates alone, the net difference in estimated and actual costs to date is only $200. This difference is so small that when a predicted total cost of the project is computed on the basis of the weighted percentage rule described above, no significant difference in the estimated total is predicted.

Table 10-2. Project Cost Report After Fourth Week (Output of System)

Report date	4
Planned cost to date	$10,000
Actual cost to date	$12,000
Planned cost of work completed	$11,800
Difference in actual and planned cost of work completed	$200 (1.7%)
Predicted total cost	$29,600 (0.00% of diff. in est.)

Summary reports such as Table 10-2 may best be provided by a computer, utilizing the input given in Table 10-1. A computer would be especially more practical as the size of the network increases to, say, more than a hundred activities. The output data may also be represented in graphical form for the review of top management. Figure 10-4 is an example.

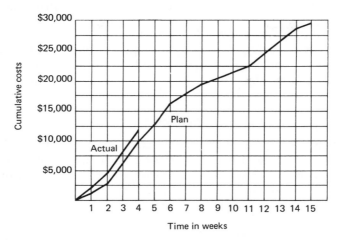

Figure 10-4 Actual expenditures plotted against the planned expenditure schedule.

OTHER SUMMARY FORMATS

Table 10-2 is only one of many cost report formats that may be designed. Another important means of summarizing the cost data is by responsibility codes. For example, all activities for which a given department or subcontractor is responsible may be coded in the input format with a particular number. Separate cost computations may then be made for the activities with the same responsibility codes. Thus, each organizational group can be given a report similar to Table 10-2 which is based on only those activities for which that group is responsible. The pottial value of breakdowns of this type is clear.

Similar codes may also be developed for activities which form a physical entity within the project, such as building A of a project involving several buildings, or the airframe of a missile development program. Summaries on these codes would cut across organizational lines to show the status of various technical subdivisions of the project.

Summarizations based on coding techniques of these types are important:

(1) to break down the total project into packages under the control of individual supervisors and subcontractors,

(2) to permit identification of components or subprojects having the most effect on the cost control problems, and

(3) in some cases, to permit the application of different overhead rates and other general expenses to different departments or subprojects.

The summarization feature is a crucial one in large organizations, for unless the data can be presented to management in a meaningful form, providing quick indications of specific cost control problems, the system will fall short of its objectives.

EXAMPLES OF OPERATING SYSTEMS

To illustrate the application of some of the basic cost control techniques presented above, several existing systems will be described. One of the systems is a "software package," made available by a computer manufacturer to users of its equipment. This system and others like it are intended to have general applicability. To other systems described in the following pages are proprietary developments by and for the using companies. One of the proprietary systems exemplifies the custom-made "PERT/Cost" systems in the aerospace industry; the other illustrates an approach taken by a construction firm. Each of the examples is described only briefly, with emphasis on the output data and report formats.

The Control Data System[1]

The PERT/Cost system for Control Data 3400/3600 series computers is essentially an implementation of the *DOD and NASA Guide.*[4] In Figure 10-5 we see the work breakdown structure, the summary and account numbers, and the relation of these elements to the detailed network.

Account codes for the system consist of summary numbers and account numbers. Summary numbers identify all blocks of the work breakdown structure except for the work packages, which are identified by account numbers. Costs are planned and collected for the work packages. Summary numbers are used to group costs for each end item subdivision on the next higher level. In addition to account or summary numbers, each block is assigned a level number corresponding to its position in the work breakdown structure; these numbers are used when requesting PERT/Cost reports. The highest level is assigned 1, with progressively higher numbers assigned to subdivision levels.

The Control Data ystem does not require costs to be estimated and collected separately for each activity on the network, but only for each work package. If costs can be defined, however, they must be represented in the network. There may be zero cost activities or joint cost

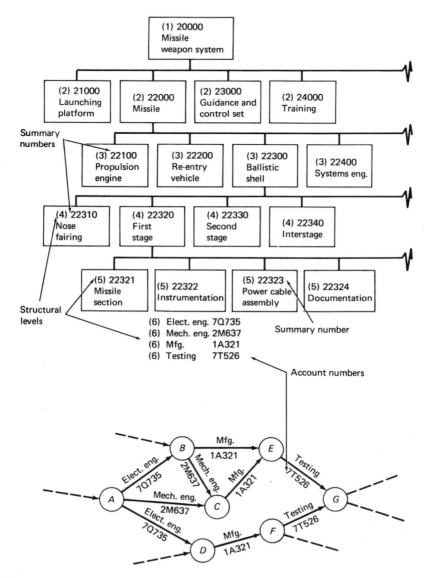

Figure 10-5 Sample work breakdown structure showing accounting identification codes. (Figures 10-5 through 10-8 *Courtesy Control Data Corporation*)

activities. A zero cost activity represents a precedence relationship only. Two or more activities may be grouped as joint cost activities when further subdivision of the work package is unwarranted. In some instances, where the activity is of greater significance, it may have individual cost, provided it is set up in a one-activity work package.

The Control Data system does not require costs to be estimated and contract, and actual. Costs are estimated by months during the project. Cost information may be specified for a span of up to 36 months. The completed estimate is the contract cost for the delivery for the project

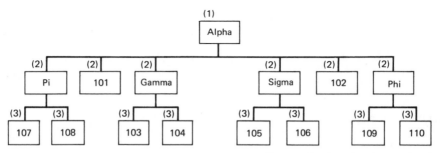

(Structural level numbers in parentheses)

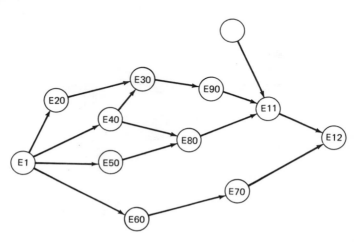

Figure 10-6 Sample work breakdown and network serving as input to illustrative reports of Control Data system.

end items. As the project progresses, actual costs are collected and processed by the system. PERT/Cost reports provide figures for value of work to date, actual cost to date, original planned total cost, and latest revised total cost. Revised costs are calculated from new forecasts which reflect re-allocation of resources. Costs may be changed at any time; they may be entered, deleted, increased, or decreased.

To illustrate how the output reports of the Control Data System may be organized, consider the sample work breakdown structure and corresponding network in Figure 10-6. In an input coding form (not shown) the network activities are assigned to various blocks and summary levels according to the nature of each activity, how it fits into a work package for cost collection purposes, and how the work packages should be related for cost summarization.

The hierachy is seen better in the sample report, Figure 10-7. Under the heading "IDENTIFICATION" the breakdown structure is given in an indented outline format. The first line of information contains the structural block description as entered on a structural block input record (here it is called "BLOCK1"). The second line under this heading contains a summary code ("ALPHA"). The summary code for the highest requested level is printed first; this code also appears after the LEVEL/SUMMARY title. The remaining entries on the report are in the form of an outline. A block at the next lower level is printed, followed by its subdivisions at the next lower level, and so forth.

The data shown for each level and work package consists of schedule and cost information. Most of the columns are self-explanatory. The cost data for Latest Revised Estimate is obtained by input changes, rather than by any internally computed projection.

A higher level report is shown in Figure 10-8. This report contains only the summary data for the two top levels of the work breakdown structure. Work packages are not shown. Note that a form of bar chart is used to display the schedule status on the right side of the report, using four letter symbols on a calendar scale.

Other similar cost-control software packages are provided by IBM Corporation, General Electric Company, and UNIVAC for users of their respective computer systems. How well such packaged systems have been integrated with the accounting procedures of existing companies is not known, since the authors are not familiar with any company that has employed one of these packages. It is likely that such packages are used primarily by companies under government contracts that require PERT/Cost reporting, and that many of these companies maintain their own internal accounting systems largely independent of the PERT/Cost system.

```
                              P E R T   S Y S T E M
                           PROGRAM/PROJECT STATUS REPORT
STAT REPORT    PROJECT      CONTRACT NO.      REPORT DATE    CONTRACT DATE   CUT OFF DATE   RELEASE DATE
                                               10/ 4/63        9/ 4/63

LEVEL / SUMMARY ITEM - 1 /  ALPHA   BLOCK1
```

CHARGE OR SUMMARY NUMBER	LEV	FIRST EVENT NO.	LAST EVENT NO.	SCHD OR ACT(A) COMPL DATE	EARLIEST -LATEST COMPL DATE	MOST CRIT SLACK (WKS)	VALUE	ACTUAL COST	(OVERRUN) UNDERRUN	PLANNED COST	LATEST REV EST.	PROJECTED (OVERRUN) UNDERRUN
												IDENTIFICATION / TIME STATUS / COST OF WORK - UNITS (WORK PERFORMED TO DATE / TOTALS AT COMPLETION)
BLOCK1 ALPHA	1	E1	E12	12/20/63	12/24/63 / 12/20/63 E12	-.40	11064	73367	(5.63) (62303)	1278651	1297118	(.01) (18467)
BLOCK2 101	2	E1	E30	A10/ 1/63	E20	-0		65840			65840	
BLOCK5 102	2	E1	E40	A10/ 1/63	E40	-0		2800			2800	
BLOCK3 GAMMA	2	E1	E30	10/23/63	10/29/63 / 10/25/63 E30	-.40	5464	2505	.54 2959	100000	75505	.24 24495
BLOCK3A 103	3	E40	E30	10/23/63	10/29/63 / 10/25/63 E30	-.40				40000	48000	(.20) (8000)
BLOCK3B 104	3	E1	E70	10/11/63	10/ 9/63 / 11/22/63 E70	6.40	5464	2505	.54 2959	60000	27505	.54 32495
BLOCK6 PHI	2	E10	E12	12/20/63	12/24/63 / 12/20/63 E12	-.40				253537	234621	.07 18916
BLOCK6A 109	3	E10	E11	11/29/63	11/25/63 / 11/29/63 E11	-.40				161737	126621	.22 35116
BLOCK6B 110	3	E11	E12	12/20/63	12/24/63 / 12/20/63 E12	-.40				91800	108000	(.18) (16200)

Figure 10-7 Cost control report of Control Data system, at work package level.

```
                P E R T   S Y S T E M
              MANAGEMENT SUMMARY REPORT
```

| STAT REPORT | PROJECT | CONTRACT NO. | REPORT DATE 10/ 4/63 | CONTRACT DATE 9/ 4/63 | CUT OFF DATE | RELEASE DATE |

LEVEL / SUMMARY ITEM - 1 / ALPHA   BLOCK1

COST OF WORK / UNITS-

ITEM	WORK PERFORMED TO DATE			TOTALS AT COMPLETION			MOST CRIT SLACK (WKS)	COMP DATE	SCHEDULE
	VALUE	ACTUAL COST	(OVERRUN) UNDERRUN	PLANNED COST	LATEST REVISED EST	PROJECTED (OVERRUN) UNDERRUN			
BLOCK1 LEVEL 1 ALPHA	11064	73367	(5.63) (62303)	1278651	1297118	(.01) (18467)	-.4	12/20/63 12/24/63 12/20/63	. S . E . L
BLOCK2 LEVEL 2 101	65840	65840			65840		-0	10/ 1/63	.A
BLOCK5 LEVEL 2 102	2800	2800			2800		-0	10/ 1/63	.A
BLOCK3 LEVEL 2 GAMMA	5464	2505	.54 2959	100000	75505	.24 24495	-.4	10/23/63 10/29/63 10/25/63	. S . E . L
BLOCK6 LEVEL 2 PHI				253537	234621	.07 18916	-.4	12/20/63 12/24/63 12/20/63	. S . E . L
BLOCK7 LEVEL 2 PI				834114	854130	(.02) (20016)	-.4	11/29/63 12/ 3/63 11/29/63	. S . E . L
BLOCK4 LEVEL 2 SIGMA	5599	2222	.60 3377	91000	64222	.29 26778	1.0	11/ 8/63 10/18/63 10/25/63	. S .E .L

SCHEDULE

S- SCHED COMPL DATE - TOTAL ITEM
A- ACTUAL COMPL DATE - TOTAL ITEM
E- EARLIEST COMPL DATE- CRIT ITEM
L- LATEST COMPL DATE- CRIT ITEM

```
P       1963        1964        5678L
YRJFMAMJJAS ONDJFMAMJJASOND   YR
```

Figure 10-8 Cost control report of Control Data system, at top summary levels.

The Hughes System†

The Hughes Aircraft Company developed its PERT Cost system in 1961 as a supplement to its basic PERT Time system. It was originally designed for operation on the IBM 7090 computer. It is a basic PERT program with several cost control and manloading report features. It does not include resource allocation or time-cost trade-off features.

In the Hughes system several levels of network summarization are defined by a work breakdown structure, which includes in order of increasing detail: program, project, task, subtask, (work package), and activity levels. One of the outstanding features of the computer pro-

Table 10-3. Principal Output Reports of the Hughes PERT Cost System†

Report	Contents	Provided For
1. Cumulative dollars	a. Estimate at completion b. Budget at completion c. Actual to date d. Estimate to date e. Remainder of budget at time now f. Estimate to complete g. Predicted balance at completion	Task, project, and program levels
2. Cumulative labor (Figure 10-9)	Same as items (a)-(g) in cumulative dollars report, except that units are man-weeks	Task, project, and program levels
3. Manpower (Figure 10-10)	Direct manpower charged per week, planned manpower per week, and equivalent number of people currently charging by organization	Task, project, and program levels
4. Cost outlook report (Figure 10-11)	Estimated, actual and budgeted costs grouped by work packages; time/cost progress ratio (per cent of estimated time expended/per cent of estimated cost expended); and exception flags, including a flag indicating a disparity between time and cost figures that exceeds a preset limit	Activity, task, project and program levels

† The program, project, and task levels referred to in this table are arbitrary designations assigned to the three levels of summary within the capability of the basic Hughes PERT System.

† Information on the Hughes system is based on reference 2 and personal correspondence with Hughes Aircraft Company management.

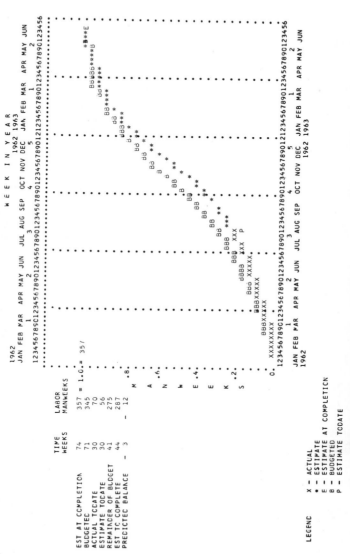

Figure 10-9 Portion of cumulative labor report produced by the Hughes PERT.
† All illustrations of this system (Figures 10-9 through 10-13) *Courtesy Hughes Aircraft Company.*

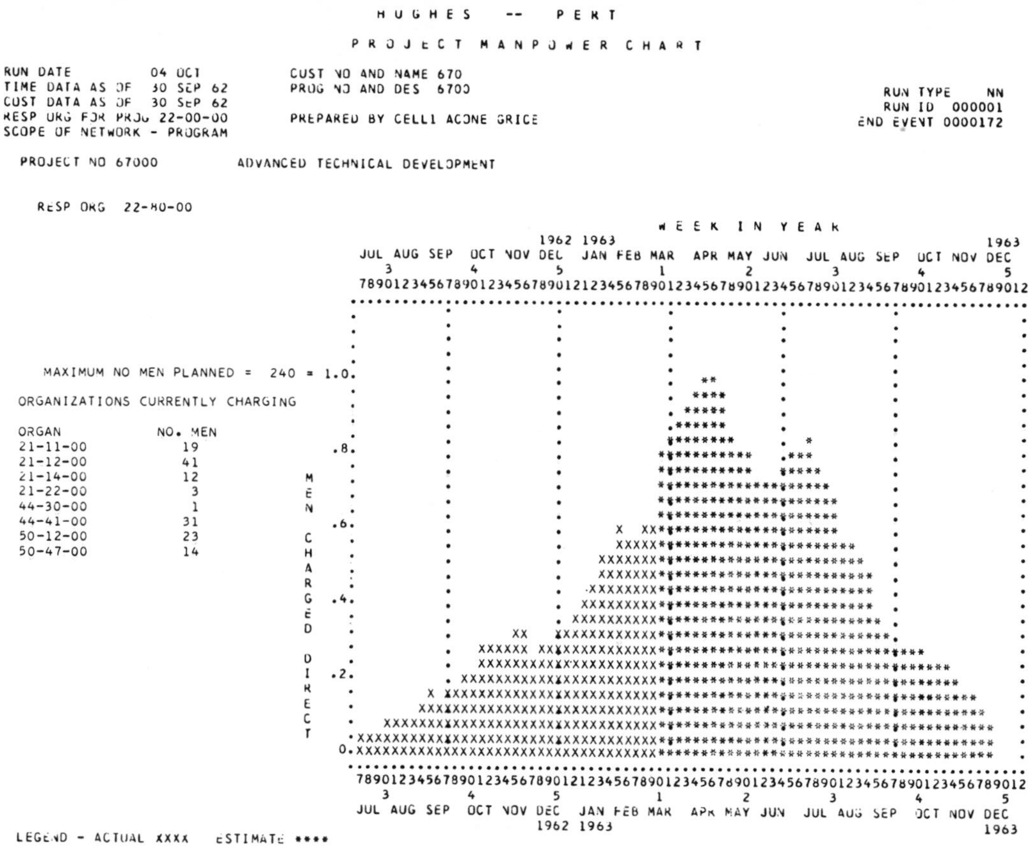

Figure 10-10 Project manpower chart, Hughes system.

```
                    H U G H E S   --   P E R T

              T A S K   R E P O R T  -  COST OUTLOOK

RUN DATE        04 OCT        CUST NO AND NAME 670          RUN TYPE   NN
TIME DATA AS OF 30 SEP 62     PROG NO AND DES 5700          RUN ID  000001
COST DATA AS OF 30 SEP 62     PROJ NO AND DES 67000         END EVENT 0000172
RESP ORG FOR PROG 22-00-00    PREPARED BY CELLI ACONE GRICE
SCOPE OF NETWORK - PROGRAM

TASK   NO 03   INITIAL PROJECT DEVELOPMENT PLAN
   CONTR    ADVANCED TECH DEV NAS5-2797
   HAC REF  9419  GLA 1326  C/A 301
   RESP ORG 22-80-XX
```

ACTIVITY DESCRIPTION ACTI NO RESP ORG PEN SEN (CHARGE NO.)	C H N G	ESTIMATED -- ACTUAL COSTS (X$1,000)								PROGRESS RATIO T/C
		LABOR EST	ACT	MAT/OTH EST	ACT	COM MIT	TOTAL EST	ACT	BDGT	

THE FOLLOWING ACTIVITIES ALL HAVE CHRG NO.1326-301-0301

```
S STRUCTURAL STUDY
2243  22-43-00
0001  0302            C   12

C FEASIBILITY REPORT
2243  22-43-00
0302  0015                 3

        SUB-TASK TOTALS *     15          15    9       80/ 60
        EST COST AT COMPLETION 12         12            E
```

THE FOLLOWING ACTIVITIES ALL HAVE CHRG NO.1326-301-0302

```
S THERMAL + STRESS STUDY
2241  22-41-00
0001  0308            C

C FEASIBILITY REPORT
2241  22-41-00
0308  0015            C

        SUB-TASK TOTALS *     30          4    34       100/ 97
        EST COST AT COMPLETION 28         5    33
```

Figure 10-11 Cost outlook report, Hughes system.

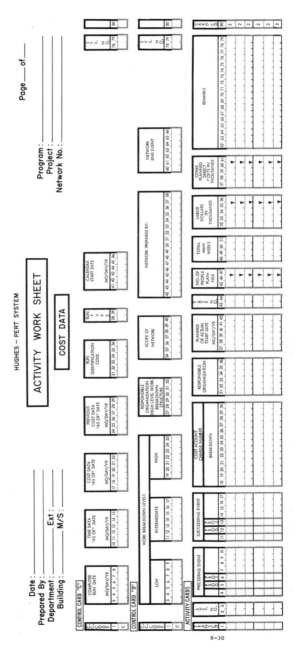

Figure 10-12 Examples of input formats for the Hughes system.

gramming for the system is the capability of the computer to summarize networks from one level to the next, making critical path and cost control computations at each level. (This feature is described in Chapter 5 under the discussion of Network Condensation.) In addition to the work breakdown structure, networks may be coded such that outputs may be summarized by (a) product end-use structure, and (b) organizational structure. These optional summaries are made possible by the crossreferencing of activities or work packages in these categories.

Among the output reports available in the Hughes system are those listed in Table 10-3, which consist of computer-produced graphic charts and tabular reports. Examples of graphic reports are shown in Figures 10-9 and 10-10. One of the tabular reports is shown in Figure 10-11. Some of the input card formats are shown in Figure 10-12.

A schematic diagram of the operation of the Hughes system is shown in Figure 10-13. Two of the unique features of the system are evident in this illustration. The first is the capability of the computer to integrate detail networks and to provide output reports associated with several summary levels. The other feature of note is at the bottom of the figure in the magnetic tape symbol labeled Cost Accts Actuals Tape. This tape is derived from the company's cost accounting system and provides the actual expenditures data in a format compatible to the PERT Cost system.

Other points of interest in the Hughes system are briefly noted below:

(1) The system has the capability of accepting either man-loading estimates (headcount per unit time) or man-hour estimates; in both modes the data are organized by activity or work package.

(2) Overhead labor costs are added to direct activity estimates by percentage rates. General and administrative expenses and fee are not included in the PERT Cost system, but may be added at the contract summary level.

The Allen Bros. & O'Hara System†

Allen Bros. & O'Hara, Inc., is a general contracting firm in Memphis, Tennessee. The cost control system the firm developed in 1958 was primarily directed toward mechanization of job estimating procedures. Later, as the firm became familiar with critical path concepts, the system was modified and expanded to provide cost control features. The system not only utilizes a computer at the company headquarters but

† Information on the Allen Bros. & O'Hara system is based on reference 3 and correspondence with Mr. Morris Beutel.

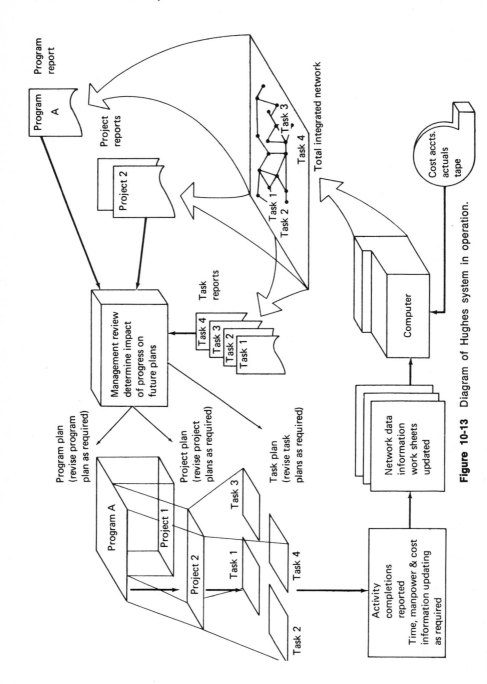

Figure 10-13 Diagram of Hughes system in operation.

also data-transmitting equipment at each job site throughout the nation; this equipment enables input data to be transmitted directly to the computer facilities by means of telephone lines.

The cost-estimating phase of the Allen Bros. & O'Hara system involves the use of an historical cost catalog, which is in the form of punched cards. The catalog maintains unit cost, compiled from previous job cost reports, by a 24-month moving average. To estimate a new job, a quantity survey is made, determining for each subdivision of the work the quantity and size of each item required by plans and specifications. These figures are fed to the computer, along with the desired crew size for each item. The computer uses these inputs with the cost catalog to produce not only an estimate of the cost of the job but also time estimates for each item. The time estimates may be used in preparing a network for the project.

The results of the estimating computations are organized into a variety of reports, among which are the following:

(1) Summary of the total job cost, in which each trade is priced separately (Figure 10-14).
(2) Unit price catalog (Figure 10-15).
(3) Separate summary for each trade (Figure 10-16).
(4) Summary sheets for subdivision of the project, including unit costs of the items involved (Figure 10-17).
(5) Weekly labor cost report (Figure 10-18).
(6) Financial status report (Figure 10-19).
(7) CPM schedule with calendar dates (Figure 10-20).

It is emphasized that the above reports are prepared prior to the beginning of work on the project. Reports (5), (6), and (7) are then updated during the course of the project. The means of updating the CPM schedule is illustrated in Figure 10-21. Using a prepunched set of cards for the activities, the superintendent of the job reports the activities completed each week by inserting the cards into data-transmitting equipment connected to the company headquarters. If an activity is only partly completed, he modifies the card to show the number of days remaining.

The financial status report is updated monthly by accounts payable and weekly cost reports, using the current updated estimated values of work as the basis of comparison. The weekly cost reports are not coded to relate directly to activities in the network, however.

Another interesting feature of the system is the routine maintenance of the historical cost catalog. The unit cost records in the catalog are updated monthly on the basis of recent cost reports. The revision compu-

NAME OF JOB - HOLIDAY UNITS
DESCRIPTION - 80 RENTAL UNITS
LOCATION - JOHNSON CITY, TENNESSEE

ALLEN BROS. & O HARA, INC.
NATIONWIDE GENERAL CONTRACTORS
3742 LAMAR AVE. MEMPHIS, TENN.

JOB NO. 1730
BLDG. SIZE 80 R-U
DATE AUG. 13, 1963

RECAP OF ESTIMATE

CODE	DESCRIPTION OF WORK	UNIT COST	FRINGE BENEFITS	F.I.C.A. INSURANCE	SALES TAX	LABOR	MATERIAL	SUB-CONTRACT	TOTAL
200	TEMPORARY CONSTRUCTION	$42.08	$.00	$73.98	$79.20	$573.50	$2,640.00	$.00	$3,366.68
300	EXCAVATION,GRADING,BACKFILL	212.75		722.47	307.68	5,733.92	10,256.09		17,020.16
600	DRAIN PIPE	10.87		16.95	21.01	131.42	700.40		869.76
900	CONCRETE WORK	396.61		337.95	836.17	2,682.16	27,872.40		31,728.68
1000	CEMENT WORK	124.62		982.42	34.66	7,796.98	1,155.23		9,969.29
1100	FORM WORK-WOOD,METAL	298.38		1,461.77	314.78	11,601.34	10,492.55		23,870.44
1200	REINFORCING RODS,MESH	179.98		352.50	327.62	2,797.60	10,920.56		14,398.28
1400	BRICK MASONRY	335.00		1,336.27	327.84	12,147.92	12,928.01		26,800.04
2200	WATER AND DAMPPROOFING	3.90		5.24	7.2	41.56	257.21		311.73
2500	STRUCTURAL STEEL & IRON	190.70		347.90	372.80	2,108.48	12,426.52		15,255.70
2800	ORNAMENTAL METAL & MISC.IRO	110.43		240.34	196.06	1,863.08	>6,535.30		8,834.78
3300	LUMBER AND ROUGH CARPENTRY	311.10		1,496.80	343.35	11,603.07	11,444.94		24,888.16

HOLIDAY INN
80 RENTAL UNITS
JOHNSON CITY, TENNESSEE

ALLEN BROS. & O HARA, INC.
MEMPHIS, TENNESSEE
UNIT PRICE CATALOG

JOB NO. 1730
AUG. 13, 1963
*IN MATERIAL COL.—SUB CONTRACT WORK

CODE NUMBER	MAT. CODE	DESCRIPTION OF WORK 1	UNIT 2	HOURS PER UNIT 3	LABOR RATE 4	LABOR AMOUNT 5	TOTAL LABOR UNIT 6	MATERIAL QUANTITY 7	MATERIAL UNIT PRICE 8	MATERIAL AMOUNT 9	TOTAL MATERIAL 10
1100.20		ERECT FOOTING & PIER FORMS	SQFT								
1100.20	A	COMMON LABOR		.014	1.250	.018					
1100.20	B	CARPENTER		.065	2.500	.163					
1100.20		CARPENTER FOREMAN		.008	2.750	.022					
1100.20	J	ERECT FOOTING & PIER FORMS	SQFT				.203			.180	.180
1100.30		WRECK FOOTING & PIER FORMS	SQFT								
1100.30	A	COMMON LABOR		.017	1.250	.021					
1100.30	B	CARPENTER		.011	2.500	.028					
1100.30	J	CARPENTER FOREMAN		.002	2.750	.006	.055				

Figure 10-14 (Above) Summary estimate of total job cost, as produced by computer in the Allen Bros. & O'Hara System.†
†All illustrations of this system (Figures 10-14 through 10-21) *Courtesy of Allen Bros. & O'Hara.*
Figure 10-15 (Below) Unit price catalog.

NAME OF JOB - HOLIDAY INN
DESCRIPTION - 80 RENTAL UNITS
LOCATION - JOHNSON CITY, TENNESSEE

ALLEN BROS. & O HARA, INC.
NATIONWIDE GENERAL CONTRACTORS
3742 LAMAR AVE. MEMPHIS, TENN.

JOB NO. 1730
BLDG. SIZE 80 R-U
DATE AUG. 13, 1963

SUMMARY OF ESTIMATE

CODE	DESCRIPTION OF WORK	UNIT	QUANTITY	LABOR UNIT	MATERIAL UNIT	SUB-UNIT	LABOR	MATERIAL	SUB-CONTRACT	TOTAL
	CONCRETE WORK									
902.10	PLACE CONCRETE DIRECT	CUYD	75	.807	13.500		$60.53	$1,012.50		$1,073.03
902.11	CONCRETE IN FOOTINGS DIRECT	CUYD	304	.807	13.500		245.33	4,104.00		4,349.33
902.12	CONC IN CURBS & GUT DIRECT	CUYD	121	.807	13.500		97.65	1,633.50		1,731.15
902.13	CONC IN SIDEWALKS DIRECT	CUYD	143	.807	13.500		115.40	1,930.50		2,045.90
902.14	CONC IN WALLS DIRECT	CUYD	2	.807	13.500		1.61	27.00		28.61
902.15	CONC IN GRADE BEAMS DIRECT	CUYD	51	.807	13.500		41.16	688.50		729.66

NAME OF JOB - HOLIDAY INN
DESCRIPTION - SITE IMPROVEMENTS
LOCATION - JOHNSON CITY, TENNESSEE

ALLEN BROS. & O HARA, INC.
NATIONWIDE GENERAL CONTRACTORS
3742 LAMAR AVE. MEMPHIS, TENN.

JOB NO. 1730
BLDG. SIZE 80 R-U
DATE AUG. 13, 1963

RECAP OF ESTIMATE

CODE	DESCRIPTION OF WORK	UNIT COST	FRINGE BENEFITS	F.I.C.A. INSURANCE	SALES TAX	LABOR	MATERIAL	SUB-CONTRACT	TOTAL
300	EXCAVATION,GRADING,BACKFILL	$143.42	$.00	$373.37	$237.00	$2,963.25	$7,899.99	$.00	$11,473.61
600	DRAIN PIPE	10.87		16.95	21.01	131.42	700.40		869.78
900	CONCRETE WORK	89.23		65.78	190.80	522.03	6,360.00		7,138.61
1000	CEMENT WORK	28.37		219.07	9.09	1,738.62	303.11		2,269.89
1100	FORM WORK-WOOD,METAL	72.15		438.30	54.04	3,478.59	1,801.43		5,772.36
1200	REINFORCING RODS,MESH	21.50		39.39	39.84	312.65	1,327.84		1,719.72
1400	BRICK MASONRY	8.14		37.44	7.96	340.34	265.22		650.96
2000	WATER AND DAMPPROOFING	2.77		1.76	6.00	14.00	200.00		221.76
2500	STRUCTURAL STEEL & IRON	17.59		22.05	36.45	133.65	1,215.00		1,407.15
2800	ORNAMENTAL METAL & MISC.IRO	8.17		15.42	15.10	119.54	503.40		653.46
3300	LUMBER AND ROUGH CARPENTRY	1.00		3.32	1.49	25.77	49.59		80.17
3500	MILLWORK & FINISH CARPENTRY	.05		.10	.10	.80	3.20		4.20

Figure 10-16 (Above) Summary of cost by trade.

Figure 10-17 (Below) Summary of cost by subdivision of the project.

ESTIMATING COPY

NAME OF JOB: HOLIDAY INN
LOCATION: BALTIMORE, MARYLAND

1689 ← JOB NO.
8/16/63 ← WEEK ENDING
25 ← REPORT NO.

ALLEN BROS. & O'HARA, INC.
MEMPHIS, TENN.
WEEKLY LABOR COST REPORT

COST THIS WEEK AMOUNT	COST THIS WEEK QUANTITY	COST THIS WEEK UNIT PRICE	CODE NUMBER	DESCRIPTION OF WORK	UNIT	QUANTITY ESTIMATE	QUANTITY TO DATE	UNIT PRICE ESTIMATE	UNIT PRICE TO DATE	UNIT PRICE DIFF.	PREMIUM TIME THIS WEEK	PREMIUM TIME TO DATE	AMOUNT ESTIMATE	AMOUNT COST TO DATE	AMOUNT VALUE AT ESTIMATED UNITS	AMOUNT DIFFERENCE
			10001	WRECK SIDEWALK	SQFT	5000		250		250			125000			
11240		—	10002	MISCELLANEOUS PROTECTION	LS								200000	36360	36360	
11240		—											325000	36360	36360	
			21000	TEMPORARY OFFICE BUILDING	BDFT	9000	6000	60	10	50			54000	6000	26000	30000
			23000	TEMPORARY TOILET FACILITIES	BDFT	2000		50		50			10000			
			24001	SIDEWALK BARRICADES	BDFT	15000	15000	69	172	103-			103500	258540	103500	155040-
			27000	JOB SIGN	EACH	1	1	15720	131910	16190-			1572	13191	1572	11619-
													169072	277731	141072	136659-
			30100	HAND EXCAVATE FOOTINGS	CUYD	239	350	3302	3729	427-			78918	130503	78918	51585-
			30101	EXCAVATE FOR CURB & GUTTERS	CUYD	110		2308		2308			25386			
			30102	EXCAVATE FOR SLAB EDGES	CUYD	21		4938		4938			10370			
			30103	EXCAVATE FOR SIGN FOUNDATN	CUYD	44		2191		2191			9640			
			30200	FINE GRADE BY HAND	SQFT	65032	30115	52	51	1		797	338166	153259	156598	3339
			30201	HAND FINE GRADE SIDEWALKS	SQFT	15335		71		71			108879			
			30301	BACKFILL AT CURB & GUTTERS	CUYD	30		1496		1496			4488			
			30302	BACKFILL AT FOUNDATIONS	CUYD	3047	290	1393	1569	176-			424464	45503	40397	5106-
			30340	SUB GRADE W P PAPER	SQ	177		350		350			6195			
			31116	MACHINE EXCAVATE FOOTINGS	CUYD	2079	450	722	632	90		61	150104	28434	32490	4056
			36650	REMOVE PILING EARTH	CUYD	1650	1650	1000	660	340			165000	108957	165000	56043
												858	1321612	466656	473403	6747
			90149	CONC IN FOOTINGS W/BUGGIES	CUYD	183	106	1620	2036	416-			29646	21582	17172	4410-
			90211	CONCRETE IN FOOTINGS DIRECT	CUYD	670	668	1291	1755	464-			86497	117259	86497	30762-
			90240	PLACE CONCRETE WITH CRANE	CUYD	2		1585		1585			317			
11302	65	205	90241	CONC IN WALLS W/ CRANE	CUYD	1031	581	1331	1317	14	201	201	137225	76496	77391	835
			90242	CONC IN BEAMS W/ CRANE	CUYD	377		1331		1331			50179			

NAME OF JOB: HOLIDAY INN HI RISE
DESCRIPTION
LOCATION: BALTIMORE, MARYLAND

CENTS NOT SHOWN
(−) = LOSS

JOB NO. 1689
DATE 7/31/63

CODE NO.	LABOR ESTIMATE	LABOR COST TO DATE	LABOR ESTIMATE TO COMPLETE	LABOR DIFFERENCE	MATERIALS ESTIMATE	MATERIALS COST TO DATE	MATERIALS ESTIMATE TO COMPLETE	MATERIALS DIFFERENCE	SUB CONTRACTS ESTIMATE	SUB CONTRACTS COST TO DATE	SUB CONTRACTS ESTIMATE TO COMPLETE	SUB CONTRACTS DIFFERENCE	DESCRIPTION	TOTALS ESTIMATE	TOTALS COST TO DATE	TOTALS ESTIMATE TO COMPLETE	TOTALS TOTAL	TOTALS DIFFERENCE	% OF COMPLETION
0100	3250	125	3124		2317			2317	19800	19800			WRECK & CLEAR	25367	19925	3124	23050	2317	785
0200	1690	2777	280	1366-	11530	1325	10205						TEMPORARY CONST	13221	4103	10485	14588	1366-	310
0300	14051	4090	9957	2	2978	1720	1257		5200	5200			EXCAVATION	22229	11011	11215	22226	2	495
0400					7280		7280		74349	74349			PILING	81629	74349	7280	81629		911
0900	19736	4757	15323	345-	128065	48428	79637						CONCRETE	147802	53186	94960	148147	345-	360
1000	37189	4934	32455	200-	3604	582	3022						CEMENT WORK	40793	5516	35477	40994	200-	135
1100	107623	33694	71244	2684	39357	26659	12698		78000	18500	59500		FORMS	224980	78853	143442	222296	2684	350
1200	62		62		102672	52664	50007		32490	16290	16200		REINF STEEL	135225	68954	66271	135225		510
1400						200		200-					MASONRY		200		200	200-	
2000									5480		5480		DAMP PROOFING	5480		5480	5480		
2300					643		643						PRECST CONCRETE	643		643	643		
2500					40475		40475		13000	2800	10200		STRUCT STEEL	53475	2800	50675	53475		052
2800	9021		9021		24730	234	24495						ORNAM IRON	33751	234	33517	33751		007

Figure 10-18 (Above) Weekly labor cost report.
Figure 10-19 (Below) Monthly financial status report.

JOB 1836 HOLIDAY INN
LOC PONTIAC, MICHIGAN

PROGRESS SCHEDULE
CRIT DENOTES CRITICAL PATH
COMPLETED & DUMMY ITEMS NOT SHOWN

TIME NOW 08/12/63
SCHEDULED FINISH DATE 01/31/64
FINISH DATE 01/31/64
AHEAD DAY,

ACTIVITY SEQUENCE		DESCRIPTION OF WORK	DAYS REQD	PREDICTED START	LATEST START	SLACK DAYS	SCHEDULED ADJUSTED PREDICTED FINISH	LATEST FINISH
1	2	50.0% C OVERLOT CUT & FILL	5.0	08/12/63	08/12/63	CRIT	08/19/63	08/19/63
1	31	50.0% FABRICATE & DEL. STR. STEEL	50.0	08/12/63	08/22/63	.0	10/22/63	11/01/63
2	3	50.0% J FTG EXCAVATIONS	1.0	08/19/63	08/19/63	CRIT	08/22/63	08/22/63
2	4	50.0% J FTG EXCAVATIONS	1.0	08/19/63	08/19/63	.5	08/22/63	08/22/63
2	12	100.0% T LAYOUT	10.0	06/19/63	10/15/63	.0	09/03/63	10/29/63
2	54	50.0% C OVERLOT CUT & FILL	5.0	08/19/63	12/19/63	18.8	08/26/63	12/27/63
2	68	100.0% UNLOAD RESTEEL	5.0	08/19/63	11/07/63	56.2	08/26/63	11/15/63
2	114	100.0% T TEMPORARY CONSTRUCTION	5.3	08/19/63	01/23/64	107.9	08/27/63	01/31/64
3	9	100.0% J FDN REINF STEEL	.9	08/20/63	08/27/63	4.2	08/21/63	08/28/63
4	9	50.0% J CONCRETE IN FOOTINGS DIRECT	.5	08/21/63	08/21/63	4.1	08/21/63	08/28/63
4	10	100.0% K FOOTING EXCAVATIONS	1.0	08/21/63	08/29/63	5.6	08/22/63	08/30/63
4	11	100.0% K CONN. DOOR FRAMES	.6	08/21/63	09/25/63	7.0	08/22/63	09/25/63
4	5	50.0% J CONCRETE IN FOOTINGS DIRECT	.5	08/21/63	08/22/63	CRIT	08/22/63	08/24/63
5	6	25.0% J MASONRY 1ST FLOOR	1.4	08/22/63	08/22/63	.0	08/24/63	08/24/63
5	9	50.0% K FTG EXCAVATION	1.0	08/22/63	08/29/63	CRIT	08/23/63	08/30/63
6	10	50.0% K CONC IN FOOTINGS	3.2	08/22/63	08/22/63	.0	08/26/63	08/26/63
6	17	100.0% K FDN REINF STEEL	1.2	08/22/63	09/27/63	5.1	08/23/63	09/30/63
6	7	50.0% L FOOTING EXCAVATIONS	.5	08/22/63	09/22/63	7.4	08/23/63	08/26/63
8	10	75.0% L RU FORMS 2ND FLOOR	3.0	08/23/63	08/21/63	.0	08/27/63	08/30/63
7	12	50.0% S EXCAV FOUNDATIONS	1.0	08/23/63	10/24/63	1.4	08/26/63	10/24/63
7	18	50.0% R CONC IN FOOTINGS	.7	08/23/63	09/30/63	5.3	08/26/63	10/01/63
7	19	50.0% R FTG EXCAVATIONS	.4	08/23/63	09/30/63	6.5	08/26/63	10/03/63
8	24	100.0% J REINF 2ND FLR SLAB	2.3	08/23/63	05/16/63	10.7	08/27/63	07/16/63
8	27	100.0% L FDN REINF STEEL	1.3	08/23/63	10/06/63	14.6	08/26/63	10/10/63
5	8	25.0% J RU FORMS 2ND FLOOR	.8	08/24/63	08/25/63	13.7	08/24/63	08/26/63
9	10	100.0% J MASONRY 1ST FLOOR	2.0	08/28/63	08/28/63	CRIT	08/30/63	08/30/63
9	24	100.0% J CONDUITS IN 2ND SLAB	3.6	08/28/63	09/12/63	10.9	09/03/63	09/16/63
9	26	100.0% S UNDERFLOOR PIPING	6.0	08/28/63	09/18/63	14.7	09/06/63	09/26/63
10	11	25.0% K MASONRY 1ST FLOOR	1.0	08/30/63	09/24/63	.0	09/03/63	09/25/63
10	17	50.0% K CONCRETE IN FOOTINGS	.5	08/30/63	09/27/63	2.5	08/30/63	09/30/63
10	18	50.0% K FTG EXCAVATION	1.2	08/30/63	09/30/63	3.6	09/03/63	10/01/63
10	24	100.0% J POUR & FIN 2ND FLOOR	12.5	08/30/63	08/30/63	CRIT	09/18/63	09/18/63
10	26	100.0% J BACKFILL AT FOUNDATIONS	1.0	08/30/63	09/25/63	17.7	09/03/63	09/26/63
12	13	50.0% S FDN FORMS	1.2	09/03/63	10/29/63	.0	09/05/63	10/31/63
12	14	100.0% C DRAINAGE EXCAVATION	1.0	09/03/63	12/06/63	.0	09/04/63	12/09/63

Figure 10-20 Critical path schedule.

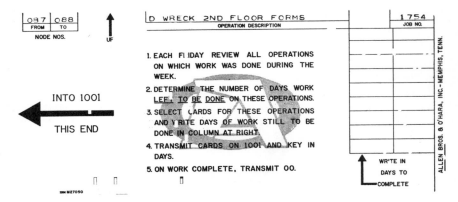

Figure 10-21   CPM weekly input for completed and partially completed activities.

tation uses a 24-month moving average incorporating a trend factor and organized by months of the year. Thus, when an estimate is prepared, the computer utilizes up-to-date unit prices, based on the season the work is to be done and the trend in costs of each item. Variations in prices due to location are also taken into account by an index for the location of the proposed project.

The computer will also estimate the time required for each bid item. For this computation it uses the data in the cost catalog and a special input that specifies the desired crew sizes for each item. The resulting time estimates may then be applied to the CPM diagram—if the company wins the contract.

SUMMARY

The concept of network cost control offers to project management a basically sound and powerful way of relating expenditures to the work done, a way of reporting what management paid for what it got. Unlike other critical path techniques, however, network cost control cannot be applied easily to only selected projects, or only during certain phases of a project. The design and implementation of such a control system involves modifications in established payroll and cost accounting procedures of the organization. To obtain an adequate return on this investment, the system should be consistently applied to the projects undertaken by the organization. Furthermore, the full use of network cost control concepts requires data processing equipment.

For these reasons, one cannot expect network cost control to receive the same rapid acceptance that basic CPM and PERT have received in their first ten years. In the aerospace industry the concept has begun to be developed and implemented at an accelerated rate, due to the interest of the Government and the general availability of data processing talent and facilities. In the other major project-oriented industry, construction, there is less incentive from the customers and fewer computers in the companies; thus, progress in this area will be less rapid. However, with the trend toward improved management controls in construction and other industries, and with the increasing economies of data processing equipment and services, there can be little doubt that network cost control systems will be developed and utilized widely in the next few years.

REFERENCES

1. Anonymous, *3400/3600 Computer Systems, PERT Reference Manual,* Control Data Corporation, Palo Alto, California, 1964.
2. Archibald, R. D., *Experience Using the PERT Cost System as a Management Tool,* presented at the Institute of Aerospace Sciences, Los Angeles, June 21, 1962.
3. Beutel, M. L., "Computer Estimates Costs, Saves Time, Money," *Engineering News-Record,* February 28 (1963).
4. *DOD and NASA Guide, PERT Cost Systems Design,* by the Office of the Secretary of Defense and the National Aeronautics and Space Administration, U.S. Government Printing Office, Washington, D.C., June, 1962.
5. *Network Analysis System,* Regulation No. 1-1-11, Department of the Army, Office of the Chief of Engineers, March 15, 1963.

EXERCISES

1. For the network in Figure 10-1, assume that it is decided to plan that all activities will begin at the *ES* times, except for the following activities, which are scheduled to begin at the times listed below:

Activity	Scheduled Start
0–1	3
1–2	5
2–5	9
0–6	4
6–7	5

Compute the planned cumulative costs for the project and plot the results on Figure 10-2.

2. Using the methods given in Chapter 8, schedule the activities in Figure 10-1 such that the project duration is minimized, subject to the constraint that the planned weekly expenditures never exceed $2,500.

3. The text lists several problems inherent in the design of network cost control systems. How are these problems approached in the Hughes system?

4. In the Allen Bros. & O'Hara System 24-month moving averages of cost data are maintained. What practical advantages would there be to using exponential smoothing in this application instead of moving averages?

11

THE PERT STATISTICAL APPROACH

In Chapter 1, PERT was described as being appropriate for scheduling and controlling projects comprised primarily of activities whose actual duration times are subject to considerable chance variation. It is because of this variability that for projects of this type, the time element of project performance is usually of paramount importance. While the CPM approach, as described in Chapters 3 and 4, is quite frequently applied to programs of this type, the single estimate, D, of the average activity performance time which it employs completely ignores the chance element associated with the conduct of the project activities. For example, an activity which is expected to take 10 days to perform, but might vary from 9 to 11 days, would be treated no differently than an activity which is also expected to take 10 days to perform, but might vary from 2 to 25 days. The advantage of the PERT statistical approach, originally developed by D. G. Malcolm[1, 2] and others, is that it offers a method of dealing with this chance variation, making it possible to allow for it in the scheduling calculations, and finally using it as a basis for computing the probability (index) that the project, or key milestones in the project, will be completed on or before their scheduled date(s). Although certain of the assumptions underlying PERT are questioned on theoretical grounds, we will describe practical procedures which can be used to circumvent these criticisms.

To give an overview of the PERT statistical approach, consider the network originally presented in Figure 4-5, and shown here in modified form in Figure 11-1. In this network, the critical path consists of the three activities 0-3, 3-7, and 7-8. Now suppose that the performance of each of these activities is subject to a considerable number of chance sources of variation such as the weather, equipment failures, personnel problems, or uncertainties in the methods or procedures to be used in carrying out the activity. It may be argued that if difficulties of one sort or another are encountered on a particular activity, that additional resources will immediately be applied to this activity, or subsequent activities on the critical path, so that the project will still be completed on time. This chapter is concerned with the problem of estimating *the probability of having to undertake such measures.* Hence, the uncertainties in performance time being referred to here are those associated with completing the originally defined activity with the originally specified resources.

Returning to the example, one notes that the actual performance times for the activities on the critical path, instead of being exactly 2, 8, and 5 days, are variables subject to random or chance variation, with mean values of 2, 8, and 5, respectively. Also, the actual time to perform the activities on the critical path is the sum of three random or chance variables, and (except for the slight possibility that activity 6-7 may be completed after activity 3-7 and activity 5-8 after activity 7-8) this sum is also the actual time to complete the project. Hence, to estimate the statistical distribution of project performance time, and in turn compute the probability of meeting a scheduled date for the completion of the project, it will be necessary to deal with the statistics of the sum of random variables. In the next section the theory of probability and statistics necessary to handle this problem will be considered. The following section will describe "conventional" PERT scheduling and probability calculation along with several practical applications. Certain embellishments to the conventional PERT procedure are given at the end of this chapter.

The treatment of PERT given in this chapter is referred to as "conventional" because the calculation of earliest and latest activity start and finish times is made in the same way as described in Chapter 4, using expected activity performance times only; the variability associated with activity performance times is involved only in the computation of PERT probabilities. In the last section of this chapter, a merge event bias procedure is given to incorporate the estimates of activity performance time variability in the scheduling computations. This procedure will correct for the ever present but usually small bias in the conventional computation of earliest times for merge point events. For example, in Figure 11-1 the (earliest) expected complete time for activity 7-8 is 15. However, since the actual complete time for activity 5-8 may be greater than that for activity 7-8, even though its expected time of 14 is less (earlier) than that for activity 7-8, the correct expected time for the completion of both activities 5-8 and 7-8 must be greater than 15. This is not unlike the problem of the expected arrival time of the last guest, all of which are invited to arrive at the same time. The expected arrival time of the last guest obviously increases as the number of invited guests increases, and does not depend solely on the guest with, say, the longest travel time. Fortunately, this merge event bias is often small. Because of the difficulty in correcting for this bias, it is neglected in most current applications of PERT. In general, the bias becomes significant at events where several activities merge, each having the same or nearly the same

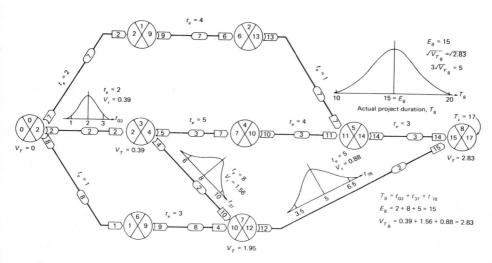

Figure 11-1 Illustration of the "conventional" PERT statistical approach to the network originally presented in Figure 4-5.

expected finish times, and considerable chance variation associated with their actual finish times.

The basic statistical concepts to be used in the development of the PERT statistical approach will be given in the next section. Readers who have a basic understanding of statistical methods should skip this section. Readers who have not had a basic course in statistics may wish to refer to a more complete treatment of this subject given in text books on statistics.[3, 4]

BASIC PROBABILITY THEORY

Probability as a Measure of Uncertainty

Some of the mathematical definitions of probability become highly abstract and the language somewhat complicated, although they will be discussed in less formidable language below. In a sense, however, one already knows what probability is all about. If one is told that an event is "almost certain," "highly probable," "about fifty-fifty," "highly unlikely," or "highly improbable," one has a good intuitive feel for the meaning of what is being said and, furthermore, this intuitive feeling is correct. All that probability theory attempts to do is to quantify these somewhat subjective statements in a precise and objective way.

In order to do this it has been found convenient to express probabilities on a scale that runs from 0 to 1. On this scale, zero represents impossibility and one represents certainty; the numbers in between represent varying degrees of likelihood. Instead of saying, for example, that it is "almost certain" that a device will continue operating for at least one more hour and "highly improbable" that it will continue operating for more than one thousand hours, one can say that the respective probabilities are, say, 0.990 and 0.001. The definitions and mathematical procedures that enable one to go from qualitative to quantitative statements can become quite technical and highly specialized; the intent, however, is to enable one to make precise and valid statements about the degree of certainty or uncertainty associated with specific occurrences.

The Managerial Function: Decision Making Under Uncertainty and Risk

The words "uncertainty" and "risk"† appear frequently in mathematical literature on probability concepts. The same two words or their synonyms

_____

† The definition of these two terms used here differs from that used in literature on decision theory, where the differentiation between uncertainty and risk is based on the presence or absence of a probability distribution associated with the variable in question.

are also a part of management's vocabulary, for the prime function of management is decision making under conditions of uncertainty with the objective of balancing the risks associated with a particular problem. Risk itself has two elements: the probability that something will happen, and the loss that will result if it does happen.

Consider the trivial example of deciding whether or not to wear a raincoat to work. If one decides to take a raincoat and it does not rain, there will be a loss, the effort or nuisance involved in carrying the raincoat. If one does not take the raincoat and it does rain; another kind of loss is involved, i.e., getting wet. The decision will, therefore, depend upon an evaluation of these possible losses and an assessment of the probability of rain. Current industrial applications of probability theory revolve around this concept of balancing risks. Probability theory is used, for example, in the determination of optimal inventory sizes, where the opposing risks are the costs of carrying too much stock, and the loss of sales that results when an out-of-stock condition occurs.

Probability has already been defined as a way of measuring uncertainty. Inasmuch as the problems facing management are the problems of uncertainty and risk, it is clear that probability has an important role to play in helping the manager to formulate and solve these problems.

The Role of Probability Theory in the PERT System

A major accomplishment of the PERT statistical procedure is the utilization of probability theory for managerial decision making. Scheduling systems have, traditionally, been based upon the idea of a fixed time for each task. In the PERT system three time estimates are obtained for each activity—an optimistic time, a pessimistic time, and a most likely time. This range of times provides a measure of the uncertainty associated with the actual time required to perform the activity sometime in the future. By means of techniques discussed below it is possible, on the basis of these estimates, to derive the probabilities that a project will be completed on or before a specified schedule date. The misleading notion of a definite time for the completion of a project, or subproject, can be replaced by statements of the possible range of times and the probabilities associated with each. The result is a meaningful and potentially useful management tool. By adding to this information an appraisal of the consequences of not meeting a scheduled date and the cost of expediting a project in various ways, management can better plan a project at the time the project proposal is prepared and at the subsequent outset of the project.

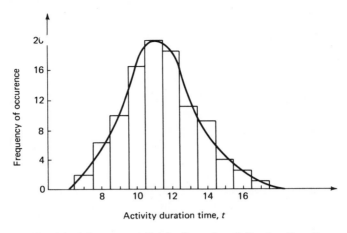

Figure 11-2 Empirical frequency distribution of activity duration times.

Empirical Frequency Distributions

To present a background in probability and statistics it is logical to begin by considering the basic raw material of statistics, i.e., observations of some measurable quantity subject to random or chance sources of variation. For discussion purposes consider a PERT activity which has been performed in the past a large number of times *under essentially the same conditions*. This assumes that no learning, changes in working conditions, job description, etc., take place. Although PERT generally involves no statistical sampling of this sort, for the purpose of this discussion, one may suppose the duration times for this activity ranged from 7 to 17 days. Now suppose that one counts the number of times the activity required 7 days to perform, 8 days to perform, etc., and displays the resulting data in the form of an empirical frequency distribution or histogram as shown in Figure 11-2. If one had an infinite number of observations and made the width of the intervals in Figure 11-2 approach zero, the distribution would merge into some smooth curve; this type of curve will be referred to as the theoretical probability density of the random variable. The total area under such a curve is made to be exactly one, so that the area under the curve between any two values of t is directly the probability that the random variable t will fall in this interval.

Characterization of an Empirical Distribution

To describe an empirical frequency distribution quantitatively, two measures are frequently employed—one which locates the point about

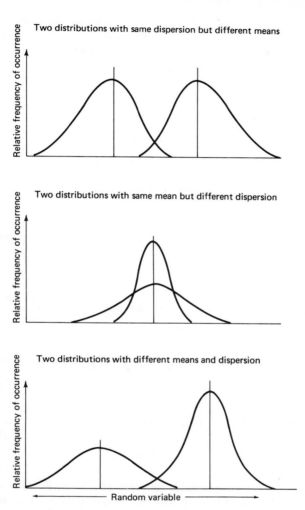

Figure 11-3 Illustration of differences in measures of central tendency and dispersion.

which the distribution is centered, a measure of its central tendency or location, and the other which indicates the spread or dispersion in the distribution, a measure of its variability. These measures are illustrated in Figure 11-3. At the top and middle of the figure, the two distributions differ either in their mean values or dispersion, while at the bottom they differ in both respects. This same information is given by quantitative measures of these two characteristics of a frequency distribution.

In PERT computations, this text will use the familiar arithmetic average or mean as a measure of central tendency, and what is called the standard deviation as the measure of variability. These statistics will first be defined with respect to a sample of n observations drawn from some distribution such as the one shown in Figure 11-2. If the n observations are denoted by $t_1, t_2, \ldots, t_n$, these measures are computed as follows:

$$\text{Measure of central tendency} = \text{arithmetic mean}$$
$$= (t_1 + t_2 + \ldots + t_n)/n = \bar{t} \quad (1)$$

Measure of variability = standard deviation = s_t

$$= \{[(t_1 - \bar{t})^2 + (t_2 - \bar{t})^2 + \ldots + (t_n - \bar{t})^2]/n\}^{1/2} \quad (2)$$

The above formula for the standard deviation indicates why it is sometimes referred to as the root-mean-square deviation; it is the square root of the mean of the squares of the deviations of the individual observations from their average. Computations will frequently use the square of the standard deviation, which, for convenience, is called the variance; $s_t^2 = $ variance of t.

Physical Interpretation of the Mean and Standard Deviation

The question usually asked at this point is what do $\bar{t}$ and s_t (or s_t^2) mean? First of all, t and s_t are estimates of the true mean and standard deviation of the distribution shown by the smooth curve in Figure 11-2. These quantities will be denoted by t_e and $(V_t)^{1/2}$, respectively; $\bar{t}$ approaches t_e and s_t approaches $(V_t)^{1/2}$ as the size of the sample, n, approaches infinity. If some assumption is made now about the theoretical distribution (the smooth curve in Figure 11-2) from which the sample was obtained, one can proceed with the interpretation. For example, suppose the random variable t is "normally" distributed, that is, the distribution has a characteristic symmetrical bell shape which frequently occurs when a variable is acted upon by a multitude of random chance causes of variation. In this case, our interpretation is shown in Figure 11-4.

Central Limit Theorem

The last bit of statistical machinery needed for PERT probability computations is the Central Limit Theorem, which is perhaps the most

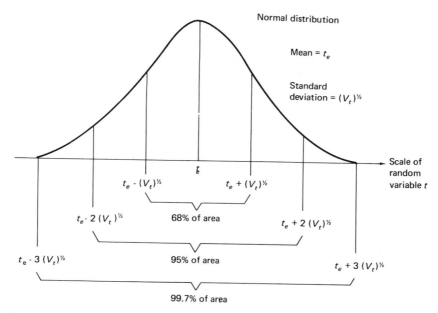

Figure 11-4   Selected areas under the normal distribution curve.

important theorem in all of mathematical statistics. In the context of PERT, this theorem may be stated in the following way.

Theorem

Suppose m independent tasks are to be performed in order; (one might think of these as the m tasks which lie on the critical path of a network). Let $t_1, t_2, \ldots, t_m$ be the times at which these tasks are actually completed. Note that these are random variables with true means $t_{e1}, t_{e2}, \ldots, t_{em}$, and true variances $V_{t1}, V_{t2}, \ldots, V_{tm}$, and these actual times are unknown until these specific tasks are actually performed. Now define T to be the sum:

$$T = t_1 + t_2 + \ldots + t_m$$

and note that T is also a random variable and thus has a distribution. The Central Limit Theorem states that if m is large, say four or more, the distribution of T is approximately normal with mean E and variance V_T given by

$$E = t_{e1} + t_{e2} + \ldots + t_{em}$$
$$V_T = V_{t_1} + V_{t_2} + \ldots + V_{t_m}$$

That is, the mean of the sum, is the sum of the means; the variance of the sum is the sum of the variances; and the distribution of the sum of activity times will be normal regardless of the shape of the distribution of actual activity performance times (such as given in Figure 11-2).

The normal distribution is extensively tabulated and therefore probability statements can be made regarding the random variable T by using these tables. A table of normal curve areas is given in Appendix 11-1, and an example illustrating its use is given later.

The Dice Tossing Experiment

To establish confidence in, and further understanding of the Central Limit Theorem, it is worthwhile to study its application to a familiar experiment—dice tossing. An experiment of tossing a single die can be described like in Figure 11-5.

If the die being tossed is unbiased, the probability of each outcome of this experiment is equally likely, and hence has a probability of 1/6, since there are just six possible outcomes. This is shown in Figure 11-6, which is called a theoretical probability distribution for the random variable X. The mean and variance of this theoretical distribution can be computed using equations (1) and (2) in which each possible value of the random variable X is weighted by its theoretical probability. These results are indicated in Figure 11-6.

$$\text{Mean of } X = (1 \times 1/6) + (2 \times 1/6) + (3 \times 1/6) + (4 \times 1/6) \\ + (5 \times 1/6) + (6 \times 1/6) = {}^{21}/_6 = 3\tfrac{1}{2}$$
$$\text{Variance of } X = (1 - 3\tfrac{1}{2})^2 \times 1/6 + (2 - 3\tfrac{1}{2})^2 \times 1/6 \\ + (3 - 3\tfrac{1}{2})^2 \times 1/6 + (4 - 3\tfrac{1}{2})^2 \times 1/6 \\ + (5 - 3\tfrac{1}{2})^2 \times 1/6 + (6 - 3\tfrac{1}{2})^2 \times 1/6 \\ = 70/24 = 2^{11}/_{12}$$

Now consider tossing two dice with the random variable Y defined as the sum of the spots on both dice, i.e., $Y = X_1 + X_2$. The probability distribution for this example, shown in Figure 11-6, follows directly from

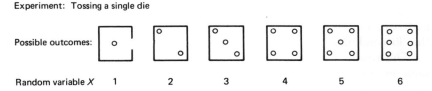

Experiment: Tossing a single die

Possible outcomes:

Random variable X   1   2   3   4   5   6

Figure 11-5

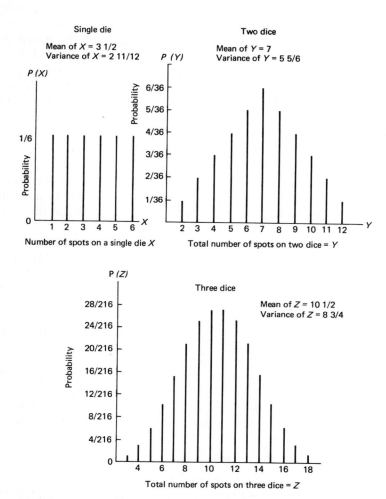

Figure 11-6 Probability distribution for tossing unbiased dice.

the fact that there are 36 mutually exclusive and equally likely ways of tossing two unbiased dice, the outcomes of which are shown by the matrix in Figure 11-7, which gives the total number of spots for each of the 36 combinations.

The reader can verify that by using equations (1) and (2) in the same manner as shown above for the single die case, the mean and variance of this distribution are exactly twice the values for a single die, i.e., mean = $2 \times 3\frac{1}{2} = 7$ and the variance = $2 \times 2\frac{11}{12} = 5\frac{5}{6}$, as they should

Second die

		1	2	3	4	5	6
	1	2	3	4	5	6	7
	2	3	4	5	6	7	8
First die	3	4	5	6	7	8	9
	4	5	6	7	8	9	10
	5	6	7	8	9	10	11
	6	7	8	9	10	11	12

Figure 11-7 The thirty-six equiprobable ways of tossing two unbiased dice.

be according to the Central Limit Theorem. One should also note from Figure 11-6 that while the basic random variable X has a rectangular distribution, the random variable Y has a triangular distribution which represents a large step toward the theoretical normal distribution as dictated by the Central Limit Theorem.

To carry this experiment still one step further, consider tossing three dice with the random variable Z defined as the sum of the spots on all three dice, i.e., $Z = X_1 + X_2 + X_3$. The reader can again verify that the mean and variance of this distribution follow the Central Limit Theorem, i.e.,

$$\text{Mean of } Z = 3 \times 3\tfrac{1}{2} = 10\tfrac{1}{2}$$
$$\text{Variance of } Z = 3 \times 2\tfrac{11}{12} = 8\tfrac{3}{4}$$

Examination of Figure 11-6 indicates that the shape of the distribution is now very close to the theoretical normal distribution. Thus, the Central Limit Theorem is illustrated, and shall be applied in the section below on the probability of meeting a scheduled date. An empirical verification of this theorem is taken up in exercise 1 at the end of this chapter.

PERT SYSTEM OF THREE TIME ESTIMATES

PERT basic scheduling computations utilize the expected values, t_e, of the hypothetical distributions of actual activity performance times, as depicted in Figures 11-2 and 11-8 below. Since PERT addresses itself

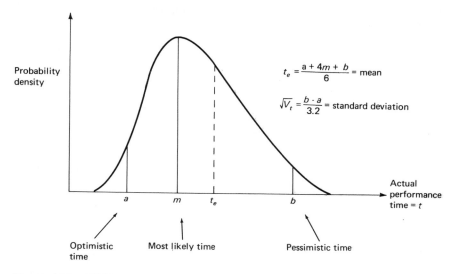

Figure 11-8 at right shows:

$$t_e = \frac{a + 4m + b}{6} = \text{mean}$$

$$\sqrt{V_t} = \frac{b - a}{3.2} = \text{standard deviation}$$

Figure 11-8 PERT system of three time estimates.

primarily to programs whose activities are subject to considerable random variation and to programs where time schedules are of the essence, it utilizes the standard deviations of the distributions shown in these figures in computing a measure of the chances of meeting various program scheduled dates.

In making PERT computations, it must be realized that the activity performance time distribution is purely hypothetical, since one ordinarily is unable to do any statistical sampling whatever. [If historical (sample) activity duration data are available, they can be used to estimate a, m, and b, as indicated in Appendix 11-2]. After an activity has been performed, the observed actual performance time, denoted by t, for the activity can be considered as a single sample from this hypothetical distribution. However, all computations are made prior to the performance of the activity; hence, as stated above, the basis of PERT computations involves no statistical sampling, but rather it depends on the judgment of the person in charge of the activity in question. The latter judgment is, of course, based on a sampling of work experiences; however, this is not sampling in the strict statistical sense. In making these estimates, one is asked to call on his general experience, and his knowledge of the requirements of the activity in question, to consider the personnel and facilities available to him, and then to estimate the three times shown in Figure 11-8. These times defined below, will then be used to estimate the mean and standard deviation of the hypothetical activity performance time distribution. The choice of a most likely time, m, and then a range of times from an optimistic estimate, a, to a pessimistic estimate, b, seems to be a natural choice of times.

Definition:

a = optimistic performance time; the time which would be bettered only one time in twenty if the activity could be performed repeatedly under the same essential conditions.

Definition:

m = most likely time; the modal value of the distribution, or the value which is likely to occur more often than any other value.

Definition:

b = pessimistic performance time; the time which would be exceeded only one time in twenty if the activity could be performed repeatedly under the same essential conditions.

The above definitions of a and b are called the 5 and 95 percentiles, respectively, of the distribution of the performance time, t. These definitions are based on a study by Moder and Rodgers.[5] They differ from the original development of PERT,[1,2] where they were assumed to be the ultimate limits, or the 0 and 100 percentiles of the distribution of t. An intuitive argument for our definition is that since the estimates of a and b are based on past experience and judgment, the 0 and 100 percentiles would be very difficult to estimate, since they would never have been experienced. Further arguments of a statistical nature will be presented below.

Although the above definitions of a, m, and b appear to be clear and workable, the following points will be helpful in obtaining *reliable* values for these time estimates.

(1) One of the important assumptions in the Central Limit Theorem is the independence of the random variables in question. Since this theorem is the basis of PERT probability computations, the estimates of a, m, and b should be obtained so that the assumption of independence is satisfied, that is, they should be made independently of what may occur in other activities in the project, which may in turn affect the availability of manpower and equipment planned for the activity in question. The estimator should

submit values for *a, m,* and *b* which are appropriate if the work is carried out with the initially assumed manpower and facilities, and under the assumed working conditions.

(2) The estimates of *a, m,* and *b* should not be influenced by the time available to complete the project, i.e., it is not logical to revise estimates by an across-the-board cut in times after learning that the project critical path is too long. This completely invalidates the PERT probabilities and destroys any positive contribution that they may be able to make in the planning function. Time estimates should be revised only when the scope of the activity is changed, or when the manpower and facilities assigned to it are changed.

(3) To maintain an atmosphere conducive to obtaining unbiased estimates of *a, m,* and *b,* it should be made clear that these are estimates and not schedule commitments in the usual sense.

(4) In general, the estimates of *a, m,* and *b* should not include allowances for events which occur so infrequently that one does not ordinarily think of them as random variables. The estimates of *a, m,* and *b* should not include allowances for acts of nature—fires, floods, hurricanes, etc.

(5) In general, the estimates of *a, m,* and *b* should include allowances for events normally classed as random variables. An example here would be the effects of weather. For activities whose performance is subject to weather conditions, it is appropriate to anticipate the time of the year when the activity will be performed and make suitable allowance for the anticipated prevailing weather in estimating *a, m,* and *b.*

Estimation of the Mean and Variance of the Activity Performance Times

It is commonly known in statistics that for unimodal distributions the standard deviation can be estimated roughly as $\frac{1}{6}$ of the range of the distribution. This follows from the fact that at least 89 percent of any distribution lies within three standard deviations of the mean, and for the normal distribution this percentage is 99.7 + percent. Hence, one can use time estimates, *a* and *b,* to estimate the standard deviation $(V_t)^{1/2}$ or the variance, V_t, as shown in equation (3):

$$(V_t)^{1/2} = (b - a)/3.2 \quad , \quad \text{or } V_t = [(b - a)/3.2]^2 \qquad (3)$$

As mentioned above, *a* and *b* were originally defined as the 0 and 100 percentiles of the distribution of *t,* and therefore, the divisor in equation (3) was 6 in place of the above value of 3.2, in the original development of PERT.[1,2] This is the basis of another argument in favor of our 5 and 95 percentile definitions of a and b. In the paper by Moder and Rodgers,[5] it is shown that the difference $(b\text{-}a)$ varies from 3.1 to 3.3 (average of 3.2) standard deviations for a wide variety of distribution types ranging from the exponential distribution to the normal distribution, including rectangular, triangular, and beta type distributions. For this same set of distributions, the difference between the 0 and 100 percentiles varies, however, from 3.5 all the way to 6.0. Thus, the use of the 5 and 95 percentiles for a and b leads to an estimator of the standard deviation that is robust to variations in the shape of the distribution of t. This is of some importance, because in general we do not know the shape of the distribution of t, and further, we wish to avoid making any specific assumptions about it.

A simple formula for estimating the mean, t_e, of the activity time distribution has also been developed. It is the simple weighted average of the estimates a, m, and b given in equation (4).

$$\text{Mean} = t_e = (a + 4m + b)/6 \tag{4}$$

To derive this formula for the mean, one must assume some functional form for the unknown distribution of t, such as shown in Figure 11-8. A likely candidate, chosen by Clark,[6] is the well known beta distribution, which has the desirable properties of being contained inside a finite interval, and can be symmetric or skew, depending on the location of the mode, m, relative to a and b. Lacking an empirical basis for choosing a specific distribution, the beta distribution was historically accepted as a mathematical model for activity duration times, for purposes of deriving equation (4) only. Using this distribution as a model and assuming that equation (3) holds,* then t_e is a cubic polynomial in m. Equation (4) is a linear approximation to the exact formula, whose accuracy is well within limits dictated by the accuracy of the estimates of a, m, and b.

It should also be pointed out that the mean is equal to the most likely or modal time $(t_e = m)$, only if the optimistic and pessimistic times are symmetrically placed about the most likely time, i.e., only if $b - m = m - a$. Thus, in the CPM procedure the single time estimate, denoted by D, is an estimate of the *mean* activity duration time and is not necessarily the *most likely* time as defined here. This is essential, since according to the Central Limit Theorem, the expected total duration of a series of activities is the sum of their mean times and not a sum of their most likely times. In fact, since the distribution of the sum

---

* The 0 and 100 percentile definitions of a and b were used in this derivation. The same formula should hold, however, for the 5 and 95 percentile definitions used in this text.

of random variables tends to the normal (symmetrical) distribution for which the mean and the mode are the same, the most likely (modal) duration time of a series of activities is not given by the sum of the individual activity most likely times, but rather by a sum of their mean times. The last statement is, of course, only approximate if the number of activities is small: however, the approximation should be very good for most practical applications to project networks.

If a single time estimate system is being used, and an activity is encountered which has a skew distribution with a considerable amount of variation, then equation (4) might be of assistance in arriving at the single time estimate, D. In this case, a person might feel that he can estimate the mean activity duration time more accurately by estimating a, m, and b, and using equation (4) to convert these numbers to the required single time estimate, D. Some evidence to this effect was found in the study cited in reference 5.

To illustrate the computation in the PERT statistical approach, consider the simple network given in Figure 11-9. Here, for the network originally presented in Figure 4-1, single time estimates have been replaced by estimates of a, m, and b. For example, $a = 1$, $m = 2$, and $b = 3$ for activity 1-2. In the middle diagram of Figure 11-9 are indicated the values of t_e, and V_t computed from equations (3) and (4). These computations are illustrated below for activity 1-2.

$$t_e = (1 + 4 \times 2 + 3)/6 = 2$$
$$V_t = [(3 - 1)/3.2]^2 = 0.391$$

The results of the forward pass are indicated by the time scale at the bottom of Figure 11-9. The earliest expected event occurrence times, E, are computed in exactly the same manner (in conventional PERT) as outlined for the single time estimate systems, as shown in Figure 4-1.*

The event variance, V_T, is computed in a manner quite similar to the computation of E. The rules are as follows:

RULE 1. V_T for the initial network event is assumed to be zero.

RULE 2. The V_T for the event succeeding the activity in question is obtained by adding the activity's variance, V_t, to the variance of the predecessor event, except at merge events.

RULE 3. At merge events, V_T, is computed along the same path used

---

* The earliest expected time for event 5, computed by the "conventional" PERT procedure, is $E = 9$. This is based on the longest path only, i.e., 1-2-3-5. If the shorter path 1-2-4-5 is also considered, using a bias correction procedure outlined later in this chapter, one obtains $E = 9\frac{1}{3}$, a small, but not entirely insignificant error. Contrary to one's intuition, the Corrected estimate of the variance, V_T, for event 5 is much smaller than the conventional estimate of 2.9.

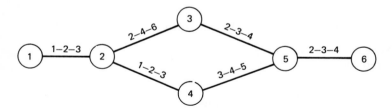

Basic network with three time estimates, *a*, *m*, and *b*.

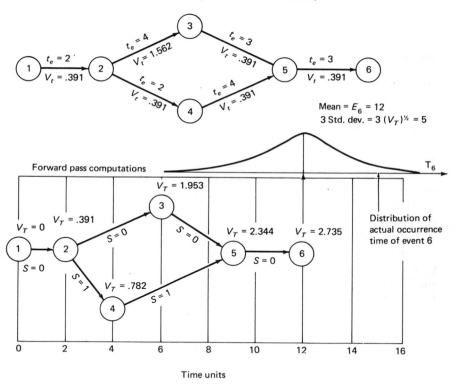

Basic network with t_e and V_t computed for each activity

Mean = E_6 = 12
3 Std. dev. = 3 $(V_T)^{1/2}$ = 5

Forward pass computations

T_6

Distribution of
actual occurrence
time of event 6

Time units

Figure 11-9   PERT statistical computations.

to obtain E, i.e., the longest path. In case of ties, choose the
path which gives the larger variance.

Applying these rules to the network given in Figure 11-9, one obtains
the following:

$$V_T \text{ (event 1)} = 0$$
$$V_T \text{ (event 2)} = 0 + 0.391 = 0.391$$
$$V_T \text{ (event 3)} = 0. 391 + 1.562 = 1.953$$
$$V_T \text{ (event 4)} = 0.391 + 0.391 = 0.782$$

Since event 5 is a merge event, its variance according to Rule 3 above is computed along the path 1-2-3-5.

$$V_T \text{ (event 5)} = 1.953 + .391 = 2.344$$
$$V_T \text{ (event 6)} = 2.344 + .391 = 2.735$$

The backward pass and slack computations are also performed in the same manner shown in Figure 4-1. For this reason, they are not given in Figure 11-9.

PROBABILITY OF MEETING A SCHEDULED DATE

Although scheduled dates could be applied to the start or finish of a project activity, they have traditionally been applied to the time of occurrence of network events. Scheduled dates are usually specified only for those events that mark a significant state in the project and vitally affect subsequent project activities; such' events are frequently called milestones. In this section, the problem of computing the probability of occurrence of an event, on or before a scheduled time, is considered.

Referring to Figure 11-9, the critical path for this network can be seen to be 1-2-3-5-6. Now consider the time to perform each of the activities along this path as independent random variables, the same assumption made during the process of collecting the a, m, and b activity time estimates. Furthermore, the sum of these random variables, which shall be denoted by T, is itself a random variable which is governed by the Central Limit Theorem. Therefore

$$T = t_{1-2} + t_{2-3} + t_{3-5} + t_{5-6}$$
$$\text{Mean of } T = E_6 = (t_e)_{1-2} + (t_e)_{2-3} + (t_e)_{3-5} + (t_e)_{5-6}$$
$$E_6 = 2 + 4 + 3 + 3 = 12$$
$$\text{Variance of } T = V_T = V_{t1-2} + V_{t2-3} + V_{t3-5} + V_{t5-6}$$
$$V_T = 0.391 + 1.562 + 0.391 + 0.391 = 2.735$$

and finally, the Central Limit Theorem enables one to assume that the shape of the distribution of T is approximately normal. This information is summarized in Figure 11-10, where the distribution is shown to "touch down" on the abscissa at three standard deviations on either side of the mean, i.e., at 12 ± 5.

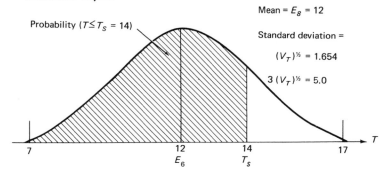

Figure 11-10   Distribution of the actual occurrence time (T) of event 6.

Now, the problem of computing the probability of meeting an arbitrary scheduled date, such as 14 shown in Figure 11-10, is quite simple. Since the total area under the normal curve is exactly one, the cross-hatched area under the normal curve is directly the probability that the actual event occurrence time, T, will be equal to or less than 14, which is the probability that the scheduled date will be met. This probability can be read from the table of normal curve areas, given in Appendix 11-1 at the end of this chapter. In order for this table to apply to any normal curve, it is based on the deviation of the scheduled date in question, T_S, from the mean of the distribution, E_6, in units of standard deviations, $(V_T)^{1/2}$. Calling this value Z, one obtains

$$Z = (T_S - E_6)/(V_T)^{1/2}$$
$$Z = (14 - 12)/1.654 = 1.21 \tag{5}$$

A value of $Z = 1.21$ indicates that the scheduled time, T_S, is 1.21 standard deviations greater than the expected time, $E_6 = 12$. Reference to Appendix 11-1B indicates that this value of Z corresponds to a probability of 0.8869, or approximately 0.89. Thus, assuming that "time now" is zero, one may expect this project to end at time 12, and the probability that it will end on or before the scheduled time of 14, *without expediting the project,* is approximately 0.89. It should be pointed out that if T_S had been two days less than E_6 instead of being greater, i.e., $T_S = 12 -2 = 10$, then $Z = -1.21$, and the corresponding probability would be 0.1131. Hence, it is essential that the correct sign is placed on the Z.

The above phrase, "without expediting," is very important. In certain projects, schedules always may be met by some means or another, for example, by changing the schedule, by changing the project requirements, by adding additional personnel or facilities, etc. The probability being computed here is the probability that the original schedule will be met *without having to expedite the work* in some way or another. For this reason, the following rules should be adopted in dealing with networks having two or more scheduled dates.

Definition:

The probability of meeting a scheduled date is the probability of occurrence of an event on or before some specified date (time).

Rule:

To compute the probability of meeting a scheduled date, the variance of the initial project event should be set equal to zero, and all scheduled dates other than the one being considered should be ignored in making the variance and probability computations.

Definition:

The conditional probability of meeting a scheduled date is the probability of the occurrence of an event on or before a specified time, assuming that all prior scheduled events occur on their scheduled dates.

Rule:

To compute the conditional probability of meeting a schedule date, set the variances of the initial project event and all scheduled events equal to zero, and then make the usual variance and probability computations.

The above definitions and rules suggest where each of these two probabilities might be applied. If one is concerned primarily with the planning of a subnetwork consisting of the activities between two scheduled events, then the conditional probability is pertinent. However, if one is concerned with the entire project, then the unconditional probability of meeting a scheduled date seems pertinent, since it gives the probability of having to expedite a project somewhere in order to meet each of the scheduled event times.

Computation of an Upper Confidence Limit on T

Suppose the scheduled time for the completion of the project shown in Figure 11-9 had been $T_s = 10$; then the probability of meeting this schedule would have been the relatively low figure of 0.11, as discussed in the previous paragraph. One might then ask the question, what revised value of T_s will increase this probability to, say 95 percent? Reference to the table of the normal curve areas given in Appendix

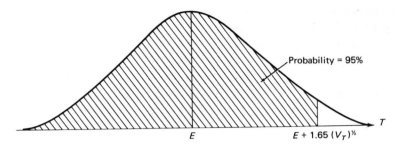

Figure 11-11   Value of **T** corresponding to an upper confidence level of 95 per cent.

11-1B indicates that 95 percent of the area is to the left of the point indicated in Figure 11-11. Hence, the answer is that if one sets $T_S = E_6 + 1.65 \ (V_T)^{1/2}$, one can be fairly confident of meeting the schedule without expediting any of the project activities. For the example given above, $T_S = 12 + 1.65 \times 1.654 = 14.7$, or 15 days. Given in Table 11-1 are formulas for T_S, corresponding to various probabilities which may be of interest in project planning.

Table 11-1.   Multiplier for Obtaining Scheduled Times With Stated Probabilities of Being Met

T_S	Probability	T_S	Probability
$E + 0.00 \ (V_T)^{1/2}$	50%	$E + 2.00 \ (V_T)^{1/2}$	97.7%
$E + 1.00 \ (V_T)^{1/2}$	84%	$E + 2.33 \ (V_T)^{1/2}$	99%
$E + 1.28 \ (V_T)^{1/2}$	90%	$E + 3.09 \ (V_T)^{1/2}$	99.9%
$E + 1.65 \ (V_T)^{1/2}$	95%		

ILLUSTRATIVE EXAMPLE USING PERCENTILE ESTIMATES

To illustrate the use of Eqs. (3) and (4), an example will be given dealing with the overhaul of jet aircraft. The data given in Table 11-2 lists the nine activities that form the critical (longest) path through the network of overhaul activities. Because of physical space limitations, the number of men that can be assigned to an overhaul task is almost fixed. Thus, an activity cannot be accelerated by adding manpower when an overhaul project is behind schedule, and for this reason, it is reasonable to assume that the actual activity duration times are statistically independent. Also, *the* critical path is considerably longer than the next most critical path, so that the usual PERT application of the Central Limit Theorem is justified. The results given in Table 11-2 indicate that the expected duration of this series of activities is 169.0 hours

(7.05 days), with a standard deviation of the actual performance times equal to 11.3 hours (0.47 days). The overhaul duration that will not be exceeded more often than say one time in ten can be estimated, using these statistics and assuming normality to be 7.65 days.

Table 11-2.   PERT Probability Analysis of an Aircraft Overhaul Project

| Activity Description | *Estimated Activity Durations (Hours)* | | | | |
	Opti-mistic	Most Likely	Pessi-mistic	Mean Time*	Variance**
Open Pylons	4	4.5	6	4.7	0.39
Open Engines	1	1.5	3	1.7	0.39
Cable Checking	10	12	16	12.3	3.53
Remove Engines	1	1	1.5	1.1	0.02
Pylon Rework	96	110	126	110.3	88.23
Reassemble Pylon, Etc.	12	16	20	16.0	6.27
Fuel Aircraft	1	2	6	2.5	2.45
Check Fuel Tanks, Etc.	4	8	12	8.0	6.27
Wing Closures	2	2.5	4	2.7	0.39
Final Checkout	6	8	20	9.7	19.03
TOTALS:				169.0	126.97
Standard Deviation = 11.3 Hours					

° Based on Eq. (3)
°° Based on Eq. (4)

In this example, the objective was to accomplish the overhaul project in seven days. Thus, technological changes in the techniques of overhaul or in the extent of the work to be accomplished must be made to reduce the overall duration of the project by at least 0.65 days, or about two work shifts, or a reduction in the standard deviation, or some combination of both.

MONITORING ACTIVITY TIME ESTIMATES AND PERFORMANCE

One of the frequent criticisms of the PERT statistical approach is that the persons supplying values of a, m, and b do not have the experience to furnish accurate data. This criticism is largely due to the nature of the work being planned. Also, the estimators are not consistent, some being conservative while others are liberal in making their estimates. In addition, there is the real possibility that the estimates are biased by a knowledge of what some higher authority would like the times to be for arbitrary reasons.

This problem was studied by MacCrimmon and Ryavec.[7] They studied the effects of various sources of errors on the estimates of the mean, t_e, and the standard deviation $(V_t)^{1/2}$. They considered errors introduced by (1) assuming the activity time distribution was a beta distribution; (2) by using the PERT approximate formulas given by equations (3)* and (4); and (3) by using estimates of a, m, and b in place of the true values. They concluded that these sources of error could cause absolute errors in estimates of t_e and $(V_t)^{1/2}$ of 30 and 15 percent of the range $(b - a)$, respectively. Since these errors are both positive and negative, however, they will tend to cancel each other.

This problem was also considered by King and Wilson,[8] who studied actual data obtained from a large scale development project involving a prime contractor and a number of subcontractors. They examined the hypothesis that there is a general increase in the accuracy of pre-activity time estimates as the beginning of the activity approaches. They rejected this hypothesis. Their data also indicated that most of the time estimates were optimistic, some being as low as 13 percent of the actual activity duration time. The same conclusion was reached in regard to ability to improve the estimates of remaining life after the activity was started. In this study, the estimate of remaining life was, on the average, 72 percent of the actual value. On the basis of these findings, the authors proposed for consideration, the upward adjustment of all time estimates. In the project studied, a multiplier of 1.39 would have reduced the average error in the activity duration times to zero.

Another approach to this problem, which these authors feel has more promise, is to work with the individuals making the time estimates to improve the future estimates by supplying them with positive feedback information. It is suggested that records be kept for each person supplying activity time estimates. These records should give the deviation of the estimated and actual activity performance times in units of standard deviations as shown in equation (6).

$$Z = (t_e = t)/(V_t)^{1/2}$$

(6)

Z = difference between estimated and actual duration time divided by the estimated standard deviation of the duration time

One problem which complicates this analysis is that the specifications for the work comprising the activity, or the level of effort applied to the activity, may be changed before the activity is completed. In these cases, the only valid procedure is to use the estimated mean

* They assumed 0 and 100 percentiles were used, and a corresponding divisor of 6 in equation (3).

time, t_e, which was made immediately after the final deviation in the activity specification or level of effort occurred.

The Z values computed for a particular estimator can then be tabulated and studied, as additional activities with which he is associated are completed. Theoretically, these Z values should vary randomly about zero from about -3 to $+3$, with the majority of the values near zero.† Deviations from this pattern have a very logical interpretation; they should be studied by the estimator who supplied the data so that they can be minimized in making future estimates.

If one classifies the estimates of the mean and variability as low, correct, or high, then there are a total of nine different combinations of these two statistics, each having its own characteristic pattern. These combinations are given in Table 11-3, and are illustrated in Figure 11-12, which was prepared from random drawings from a normal distribution with means of -3, 0, or $+3$ when the estimated mean was low, correct, or high, respectively, and with a standard deviation of 2, 1, or 1/2 when the standard deviation was estimated to be low, correct, or high respectively. Note that the result of underestimating the standard deviation is a pattern of Z values which shows excessive dispersion.

Table 11-3. Nine Possible Patterns of Z Values in Monitoring Activity Time Estimates

Estimated Standard Deviation	Estimated Mean	Pattern Number	Method of Correction
low	low	1	increase difference, $b - a$; increase level of a, m, and b
low	correct	2	increase difference, $b - a$
low	high	3	increase difference, $b - a$; reduce level of a, m, and b
correct	low	4	increase level of a, m, and b
correct	correct	5	**desired pattern—no correction needed**
correct	high	6	reduce level of a, m, and b
high	low	7	reduce difference, $b - a$; increase level of a, m, and b
high	correct	8	reduce difference, $b - a$
high	high	9	reduce difference $b - a$; reduce level of a, m, and b.

† Because of the possible skewness in the hypothetical activity performance time distribution, Z could vary over a wider range, say -4 to $+4$. However, if we are only considering a relatively small sample of Z values for a particular individual, say 5 to 10 values, then the occurrence of these extreme values is unlikely. The range -3 to $+3$ seems to be a workable compromise.

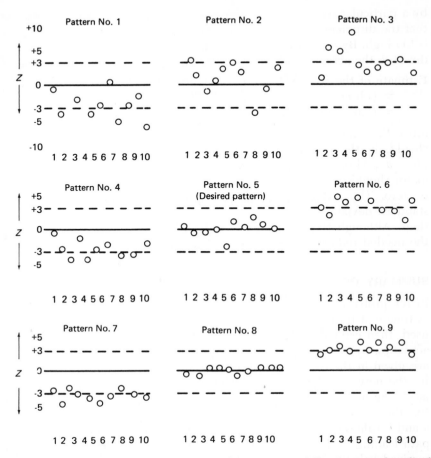

Figure 11-12 Illustration of the nine possible patterns of Z values as described in Table 11-3.

In the first three patterns of Figure 11-12, the estimate of the standard deviation is low, and hence the pattern of points shows a greater dispersion than expected. For patterns 4, 5, and 6, the estimate is correct, and the resulting dispersion of the Z values illustrates the correct pattern. For the last three patterns, the estimate is too high, and hence the pattern of Z values exhibits less dispersion than expected. Similarly, in the first column of patterns, the estimate of the mean is obviously low, in the middle column it is correct, and in the right hand column it is high.

Now suppose Pattern No. 3 is obtained from the Z values generated by a particular estimator. A visual inspection of these Z values indicates that the dispersion of the points is too great and the mean of the points is too high. By using the constants given in Table 11-7, found in Appendix 11-2, the actual standard deviation of the Z values can be estimated.

Estimate of the standard deviation of the Z *values*
$$= \text{(observed range of Z values)}/(d_2 \text{ for sample size} = 10)$$
$$= (7.5 - 1)/3.08 = 2.1$$

Since the standard deviation of Z should be about one, this estimator should, in the future, double his current estimates of the difference $(b - a)$. Also, since this pattern of points averages about +3, this estimator should decrease his estimates of a, m, and b to reduce this mean to zero. The reduction should be equivalent to about three current standard deviations, or one and one-half of the revised standard deviations. Although this is not a completely satisfactory way of expressing the needed reduction, it can serve as a guide.

SUMMARY OF HAND PROBABILITY COMPUTATIONS

PERT probability computations can be handled in much the same way as time-cost tradeoffs were handled in Chapter 9. If a computer is being used, then estimates of a, m, and b must be obtained and used on all activities. In this case, probabilities of meeting specified scheduled times are given in the computer output, such as are shown in Figure 11-13b, for the network shown in Figure 11-13a. However, if the computations are being made by hand, one could start off with a single time estimate for the mean performance time for each activity, and then obtain a and b values only for those activities on the critical path, or the longest path leading to the scheduled event in question. This procedure is summarized below.

(1) Make the usual forward and backward pass computations based on a single-time estimate, D, for each activity.

(2) Suppose one wishes to compute the probability of meeting a specified scheduled time for event X. Then obtain estimates of a and b for only those activities that comprise the "longest path" from the initial event to the event X.

(3) Compute the variance for event $X,(V_T)$, by summing the variances for the activities listed in Step 2. V_T = sum of values of $[(b - a)/3.2]^2$, for each activity on the "longest path" leading to event X.

(4) Compute Z using equation (5) and look up the corresponding probability in the normal curve table given in Appendix 11-1.

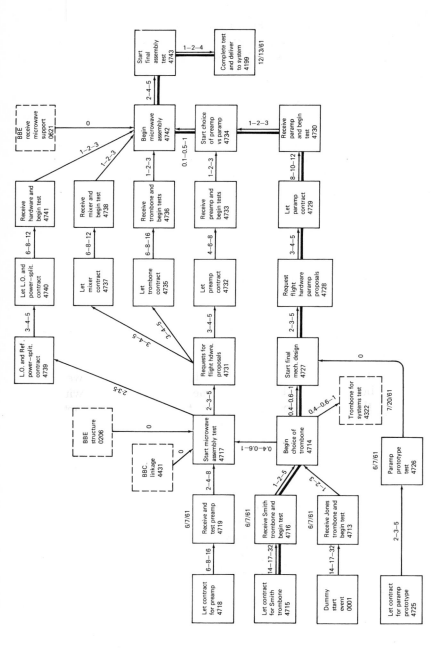

Figure 11-13a Typical PERT network of an electronic module development project. (*Courtesy Applied Physics Laboratory, John Hopkins University*)

PERT SYSTEM
PERT SYSTEM

PAGE 1

RUN 1
BY PATHS OF CRITICALITY
CHART AJ   LR SN 9 ELECTRONIC MODULE (ILLUSTRATIVE NETWORK)   SYSTEM W 034   DATE 06-07-61

EVENT PREDECESSOR	SUCCESSOR	NOMENCLATURE	DEP	DATE EXPECTED	ENDING EVENT ALLOWED	DATE SCHD/ACT	PROB	SLACK	EXP TIME	EXP VAR
4004-715	4004-716	REV DATE (SMITH TROMBONE RECD-BEG TEST)	98		05-26-61	A06-07-61		-1.6	+	
4004-716	4004-714	SMITH TROMBONE TESTED	0146	06-23-61	06-12-61			-1.6	+ 2.3	.4
4004-714	4004-727	TROMBONE CHOSEN-BEGIN MECH DESIGN	0146	06-28-61	06-16-61			-1.6	+ 3.0	.5
4004-727	4004-728	RFP PARAMP FLIGHT HARDWARE	0146	07-20-61	07-08-61			-1.6	+ 6.1	.7
4004-728	4004-729	PARAMP CONTRACT LET	0146	08-17-61	08-05-61			-1.6	+10.1	.8
4004-729	4004-730	PARAMP RECEIVED		10-26-61	10-14-61			-1.6	+20.1	1.3
4004-730	4004-734	PARAMP TESTED	0146	11-09-61	10-28-61			-1.6	+22.1	1.4
4004-734	4004-742	CHOICE BETWEEN PREAMP-PARAMP	0146	11-13-61	11-01-61			-1.6	+22.6	1.4
4004-742	4004-743	COMPL MICROWAVE ASSY	0146	12-09-61	11-28-61			-1.6	+26.5	1.6
4004-743	4004-199	COMPL FINAL TEST MICWAVE ASSY-DELIVERED	0146	12-25-61	12-13-61	12-13-61	.12	-1.6	+28.6	1.9
4000-001	4004-713	REV DATE (JONES TROMBONE RECD-BEG TEST)	99		05-29-61	A06-07-61		-1.3	+	
4004-713	4004-714	JONES TROMBONE TESTED	0146	06-21-61	06-12-61			-1.3	+ 2.0	.1
4004-714	4004-717	TROMBONE CHOSEN-BEGIN MICWAVE ASSY TEST	0146	06-28-61	07-02-61			+ .5	+ 3.0	.5
4004-717	4004-731	RFP FOR FLIGHT HDW-MIXER-TROMB-PREAMP	0146	07-20-61	07-24-61			+ .5	+ 6.1	.7
4004-717	4004-739	COMPL MICWAVE ASSY TEST-RFP LOC OSCIL	0146	07-20-61	07-24-61			+ .5	+ 6.1	.7
4004-731	4004-735	TROMBONE CONTRACT LET	0146	08-17-61	08-21-61			+ .5	+10.1	.8
4004-731	4004-737	MIXER CONTRACT LET	0146	08-17-61	08-21-61			+ .5	+10.1	.8
4004-739	4004-740	CONTRACT LET FOR LOC OSCIL AND PWR SPLT	0146	08-17-61	08-21-61			+ .5	+10.1	.8
4004-735	4004-736	TROMBONE RECEIVED		10-14-61	10-18-61			+ .5	+18.5	1.8
4004-737	4004-738	MIXER RECEIVED		10-14-61	10-18-61			+ .5	+18.5	1.8
4004-740	4004-741	LOC OSC-PWR SPLITTER RECEIVED		10-14-61	10-18-61			+ .5	+18.5	1.8
4004-736	4004-742	TROMBONE TESTED	0146	10-28-61	11-01-61			+ .5	+20.5	1.9
4004-738	4004-742	MIXER TESTED	0146	10-28-61	11-01-61			+ .5	+20.5	1.9
4004-741	4004-742	LOC OSC-PWR SPLITTER TESTED	0146	10-28-61	11-01-61			+ .5	+20.5	1.9

Figure 11-13b Typical PERT computer output. (First three-paths shown here)

INVESTIGATION OF THE MERGE EVENT BIAS PROBLEM

As pointed out in the introduction to this chapter, the conventional PERT procedure described above always leads to an optimistically biased estimate of the earliest (expected) occurrence time for the network events. This bias arises because all subcritical paths are ignored in making the forward pass computations. If the longer path leading to a merge event is much longer than the second longest path, and/or the variance of the activities on the longest path is small, this bias will be insignificant. The first part of this section will be devoted to a series of examples designed to give the reader an appreciation for the signicance of this bias problem. A brief review of the literature dealing with analytical solutions to this problem will follow. Finally, a solution to this problem using Monte Carlo simulation will be described in some detail. This approach to the problem appears to hold the greatest promise at the present time.

Magnitude of Bias

A study of this problem was made by MacCrimmon and Ryavec,[7] who considered two of the more important factors affecting the magnitude of the merge event bias. First, one would intuitively expect the bias to increase as the number of parallel paths to the network end event increases. This is studied in Figure 11-14 below. Second, one would also expect the bias to increase as the expected length of the parallel paths become equal. This is studied in Figure 11-15.

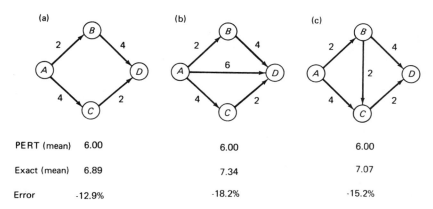

	(a)	(b)	(c)
PERT (mean)	6.00	6.00	6.00
Exact (mean)	6.89	7.34	7.07
Error	-12.9%	-18.2%	-15.2%

Figure 11-14 Effect of parallel paths, with and without correlation, on the merge event bias.

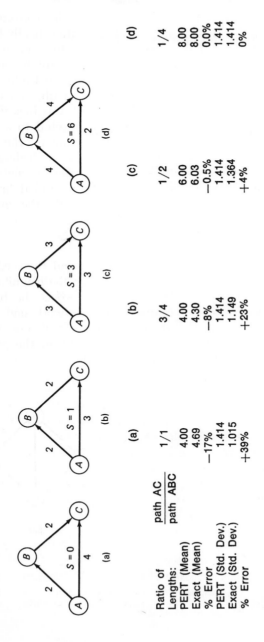

Ratio of Lengths: $\dfrac{\text{path AC}}{\text{path ABC}}$	(a) 1/1	(b) 3/4	(c) 1/2	(d) 1/4
PERT (Mean)	4.00	4.00	6.00	8.00
Exact (Mean)	4.69	4.30	6.03	8.00
% Error	−17%	−8%	−0.5%	0.0%
PERT (Std. Dev.)	1.414	1.414	1.414	1.414
Exact (Std. Dev.)	1.015	1.149	1.364	1.414
% Error	+39%	+23%	+4%	0%

Figure 11-15. Effect of Slack on Merge Event Bias.

Consider the four activity network in Figure 11-14a. The particular discrete distribution assumed for each of these activities can be identified in Table 11-4 by the corresponding mean shown on the network activi-

Table 11-4.   Discrete Distribution for Activities in Figure 11-14.

t	Probability		t	Probability
1	¼		2	¼
2	½		4	½
3	¼		6	¼
Mean $= t_e$	$= 2$		Mean $= t_e$	$= 4$
Std. Dev. $= (V_t)^{1/2}$	$= 0.707$		Std. Dev. $= (V_t)^{1/2}$	$= 1.414$
Coef. of Car. $= (V_t)^{1/2}/t_e$	$= 35\%$		Coef. of Car. $= (V_t)^{1/2}/t_e$	$= 35\%$

ties. There are two paths, *ABD* and *ACD*, both having a mean length of 6. The mean of the maximum time distribution, or the earliest expected time for event *D* is 6.89. Thus, the error in the PERT calculated mean is 12.9 percent of the actual mean.

There are two possible ways a third path, with a mean length of 6, may be created by adding one more activity. In one case the path may be completely independent of the other two paths, thus resulting in a third parallel element, *AD*, as depicted in Figure 11-14b. Alternatively, an activity *BC* can be added with a mean time of 2, thus creating path *ABCD*, shown in Figure 11-14c. In both cases there are three paths, all of mean length 6, and the network has four events and five activities.

The addition of the third path in parallel leads to an increase in the deviation of the PERT-calculated mean (still 6) from the actual mean, which in this case is 7.336. Thus, the error has increased to 18.2 percent. Figure 11-14c, on the other hand, is a network configuration, where there is a cross connection between two parrallel paths. Since there are three paths, one would expect a larger error than in a similar network with only two paths (such as Figure 11-14a), although not as large an error as in Figure 11-14b, where the three paths are in parallel. The correlation (resulting from the common activities) in the network of Figure 11-14c does indeed have the effect discussed, and the mean of the maximum time distribution lies between these two bounds, being 7.074. The error as a percent of the actual mean is 15.2 percent.

The examples given above are extreme cases, since all the paths have the same expected duration—hence, they are all critical paths. If the durations of some paths are shorter than the duration of the longest path, their effect on the project mean and standard deviation would not be as great. However, if they have a mean duration very close to the mean duration of the critical path, they would not be critical but they would have an effect almost as significant as the examples of the previous sections. The following examples shown in Figure 11-15 indicate the effect of slack in a path length.

The simple network has only two paths, *ABC* and *AC*. All activities are assumed to be normally distributed, with standard deviation equal to 1, and the appropriate mean given on the diagram. It may be noted from the diagrams that various lengths were assumed for paths *ABC* and *AC*, ranging from both of them being of equal length, to path *AC* being only ¼ the length of path *ABC*.

This example indicates that the deviation of the PERT-calculated mean and standard deviation, from the actual mean and standard deviation, may be quite large when the paths are about equal in length, but the difference decreases substantially as the path lengths become farther apart.

Another study similar to the above was made by Klingel.[9] He considered a network comprised of multiple restaurant-service station installations. Ten installations were diagrammed in parallel, including elements from market research, site selection, property surveys, zoning requirements, etc., through construction, hiring of personnel, installation of equipment, to actual opening for business. Each installation required above 100 activities. In addition, about 100 activities common to all restaurants tied the ten parallel installations together with common constraints. These comprised such items as ordering consumables, warehousing systems, developing accounting procedures, advertising campaigns, etc. Thus, the entire network contained about 1100 activities.

A simulation experiment was then conducted, using the number of parallel installations as one variable and the coefficient of variation for the individual activity duration times as a second variable. The latter was fixed at 0, ⅙, ⅓ and ½, which later proved to be a good choice, since field experience on actual duration times indicated the coefficient

Table 11-5. Averages of Simulated Project Durations

No. of Parallel Installations	Project Completion Time in Days.* Coefficient of variation $(V_t)^{1/2}/t_e$			
	0	1/6	1/3	1/2
1	228	228.64	241.36	241.04
2	228	237.36	257.32	278.68
3	228	245.64	262.32	285.48
5	228	241.08	277.3	310.32
8	228	246.28	283.04	315
10	228	252.4	285.96	334.44

* All entries are the mean of 25 replications, i.e., 25 Monte Carlo simulations, individual runs not being given.

of variation between ⅓ and ½ for the entire project. The results for the 24 combinations of variables are given in Table 11-5. To obtain an appreciation for the degree of variation in activity duration times represented by the assumed levels of the coefficient of variation, Table 11-6 has been prepared. It gives the values of a, m, and b for an activity with a mean of 25 and the indicated levels of the coefficient of variation.

Table 11-6. Representative Values of *a*, *m*, and *b* for Specified Values of the Coefficient of Variation

CV*	a	m	b	Mean	Standard Deviation
0	25.0	25.0	25.0	25.0	0.0
1/6	18.3	25.0	31.7	25.0	4.2
1/3	11.7	25.0	38.3	25.0	8.3
1/2	5.0	25.0	45.0	25.0	12.5

* CV = Coefficient of Variation = Std. Dev./Mean.

To obtain the results given in Table 11-5, each activity duration time was assumed to be normally distributed, with a mean value estimated by the project personnel, and a variance determined by the assumed level of the coefficient of variation shown in Table 11-6. Random samples from these distributions were then obtained by generating them on the computer, using an appropriate mathematical procedure such as can be found in the text by McMillan and Gonzales.[10] Having generated a time for each activity, the basic (deterministic) forward pass calculations (Chapter 4) were made to determine the total project duration time. Each of the values given in Table 11-5 are the averages of 25 such simulations of the complete project.

The results shown in Table 11-5 indicate that actual project duration times can run as high as 50 percent greater than the PERT estimate of 288 working days. Although this is an extreme value, the other errors indicated in this table are appreciable.

Rules of Thumb on Merge Point Bias

To summarize qualitatively the above results on merge event bias, it can be noted that the magnitude of the bias correction at a given merge event increases as

(1) the number of merging activities increases,
(2) the expected complete times of the merging activities get closer together,
(3) the variances of the merging activities increase, and

(4) the correlation among the merging activity complete times approaches zero.

Because of point 2 above, the correction at most merge events will be negligible and thus can be ignored. From a study of tables derived by Clark,[11] giving the expected value of the greatest of a finite set of random variables, this can be stated as a useful rule of thumb as follows.

Rule:

If the difference between the expected complete times of the two merging activities being considered is greater than the larger of their respective standard deviations, then the bias correction will be small; if the difference is greater than two standard deviations, the bias will be less than a few percent and can be ignored. (The difference referred to here is what has been defined in Chapter 4 as activity free slack.) If there are more than two merging activities, this rule should be applied to the two with the latest expected finish times.

The validity of this rule is illustrated in Figure 11-15. The difference (slack) in the expected time of the two merging activities is less than one standard deviation in Figure 11-15b and is greater than two in Figure 11-15c. The corresponding biases of 8 percent and 0.5 percent are appreciable and insignificant, respectively, as suggested by this rule.

If the above does not rule out the need for a bias correction, then it should be made. The next section reviews the studies giving analytical procedures to correct for merge event bias. This is followed by a recommended Monte Carlo correction procedure.

Analytical Merge Event Bias Correction Procedures

The merge event bias correction problem is essentially a statistical problem, dealing with a random variable defined as the maximum value of a set of random variables, not necessarily statistically independent. The latter condition complicates the problem greatly. The maximum value is the earliest expected occurrence time of the (merge) event in question, and the set of random variables is the actual complete times of the activities merging to the event in question. These latter times are not always statistically independent, because of the network cross-over condition previously illustrated in Figure 11-14c.

This is an intriguing statistical problem that has caught the fancy of many researchers. The work of Clark (1961),[11] who was a member of the original PERT development team, has been cited above. Moder (1964)[12] incorporated Clark's work in a procedure given in Appendix 9-5 of the first edition of this text; this study was also the basis of the rule of thumb given above. Fulkerson (1962)[13] studied this problem and developed a method of getting a fairly good lower bound on the true merge event occurrence time. His work was based on the assumption that each activity has a discrete probability distribution, such as was shown in Table 11-4 above. Clingen (1964)[14] extended this work to include the case when the activity duration times were assumed to be continuous. Elmaghraby (1967)[15] developed two approaches to improve on Fulkerson's method.

A rather different approach to the problem was taken by Charnes and Cooper (1964),[16] who studied the problem from a stochastic linear programming point of view. (Charnes' deterministic linear programming approach is discussed in the Appendix to Chapter 9 of this text). Still others, including Martin (1965)[17] and Hartley and Wortham (1966)[18] have approached this problem from a basic statistical distribution point of view. The work of Elmaghraby and Pritsker (1966), described in the section on Generalized Networks in Chapter 6, is also applicable to this program. (Using their nomenclature, PERT events are of the *and* type.)

There is, indeed, a wide array of approaches that have been developed to solve this problem, all of which require the use of computers for practical-sized problems. It is difficult to say which is the best procedure for a particular application, because each requires different basic assumptions and input data, different amounts of computing, and each gives different accuracy in the final estimates. It is the authors' opinion, considering all factors, that the current most economical solution to this problem is via Monte Carlo simulation.

Monte Carlo Simulation Approach to Merge Event Bias Problem

The Monte Carlo simulation approach to the solution of this problem was used by Klingel (1966)[9] in the study described above. An earlier study by Van Slyke (1963)[19] treated the methodology of this approach to the problem, which he recognized as the problem of simply solving the network model to find something that corresponds, in some sense, to the project duration and critical path in the deterministic case. The difficulty here was avoided to some extent by approximating the random problem by a series of problems of the deterministic form. To accomplish this, Monte Carlo simulation was used. A bonus from this approach was that it not only gave unbiased estimates of the mean and variance of the project duration, along with the distribution of total project time, but it also gave estimates for quantities not obtainable from the standard PERT approach. In particular, the 'criticality' of an activity, i.e., the probability of an activity being on the critical path, can be calculated. One of the more misleading aspects of conventional PERT methods is the implication that there is a unique critical path. In general, any of a number of paths could be critical, depending on the particular realization of the random activity durations that actually occur. Thus, it makes sense to talk about a criticality index. This appears to be an exceedingly useful measure of the degree of attention an activity should receive by management, and is not as misleading as the critical path concept used in PERT. It should be added that the probability of an activity being on the critical path is not correlated too well with slack, as computed by the conventional PERT procedure, which is the factor that usually determines the degree of attention that a particular activity receives.

The Monte Carlo simulation procedure was applied by Van Slyke to the network given in Figure 11-16. Each activity was assumed to have a beta distribution with mean, t_e, and a variance, V_t, as noted on each activity. As generally recommended by Van Slyke for this purpose, 10,000 sets of random times were generated for each activity in the network. For each of these sets, the longest path through the network was determined: its duration was noted, as well as a count for each activity on the critical path. The results of these 10,000 simulations are given in Figure 11-17, where the probability that an activity was on the critical path is noted on each activity. For example, 0.737 on activity 12 means that in 7370 of the 10,000 simulations, this activity was on the longest path in the network. Also given at the bottom of Figure 11-17 are the statistics pertaining to the total project duration. We note here that the PERT estimate of the project mean was low (optimistic) by only 1.5 percent; however, the variance was estimated too high by 45

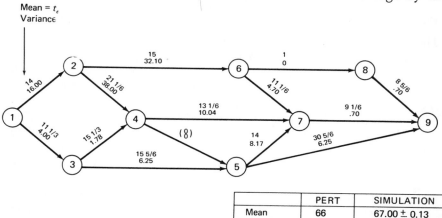

	PERT	SIMULATION
Mean	66	67.00 ± 0.13
Variance	60.27	42.39 ± 2

Figure 11-16   Illustrative network for Monte Carlo simulation.

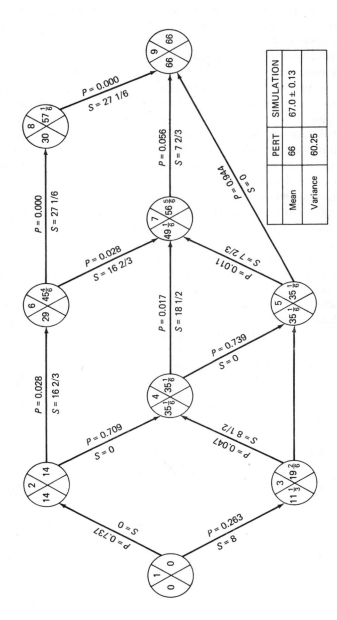

Figure 11-17 Activity criticality and project duration statistics determined by Monte Carlo simulation.

percent. The former result is about as expected according to the rule given in the previous section, because there is a considerable amount of slack along the subcritical path at each merge event. It is interesting to note, however, that activities 13 and 35, which are not on the conventional PERT critical path, have appreciable probability of ending up on the actual critical path.

Another output of the Monte Carlo simulation study is given in Figure 11-18, where the cumulative probability of a specified project duration is given, and is compared with the results given by the conventional PERT procedure based on the Central Limit Theorem.

A final note on this procedure regards the cost of obtaining these results. As an example, 10,000 simulations of a 200 activity network required 20 minutes on an IBM 7090 computer. The accuracy of the estimates based on 10,000 simulations is noted on Figure 11-17 by the intervals on the mean, i.e., 67.00 ± 0.13. This is a 95 percent confidence interval on the true (unknown) mean project duration. In most cases this degree of accuracy, i.e., ± 0.13, could be relaxed and thereby the 10,000 simulations reduced to perhaps as low as 1000. Also, if a third generation computer is used to carry out the computations, the run time could be further reduced by a factor of $\frac{1}{5}$ to $\frac{1}{10}$. Thus, it is entirely feasible to analyze networks that are of any size likely to be encountered in practice by this Monte Carlo simulation procedure.

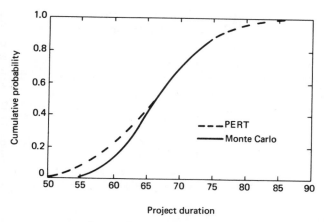

Figure 11-18 Cumulative probability of project duration.

SUMMARY

In this chapter, the PERT statistical approach to project planning and control was given, which leads to a probability that a given scheduled event occurrence time will be met, *without having to expedite the project.* The conventional PERT procedure derives its measure of uncertainty in the event occurrence times from the three performance time estimates, optimistic, pessimistic, and most likely, for each network activity. This procedure was modified by defining the optimistic and pessimistic times as 5 and 95 percentiles, respectively, of the hypothetical activity performance time distribution, rather than the end points of the distribution. Based on the Central Limit Theorem, estimates of the mean and variance in activity performance times were then used to compute a probability of meeting arbitrary scheduled times for special network events. It was recognized that it is difficult to obtain accurate estimates of the activity performance times, and procedures for improving the estimation by feedback of past estimation performance was outlined.

The merge event bias, introduced in conventional PERT by ignoring all but the critical path, was then discussed with examples illustrating the magnitude of this problem. A simple rule of thumb was given to determine, from the completed conventional PERT analysis, whether this bias will be serious or not. This rule simply states that at merge events, one can ignore the activities with free slack, if the latter is greater than one or two standard deviations of the activity completion time along the longest path.

Analytical procedures to remove the merge event bias were then reviewed, with the conclusion that the most practical procedure, at present, is to apply Monte Carlo simulation, when the above rule of thumb indicates that corrections should be made. This procedure was described and illustrated, and it was pointed out that on third generation computers, this solution to the problem is economically feasible on any size network that might be encountered in practice.

REFERENCES

1. *PERT, Program Evaluation Research Task, Phase 1 Summary Report,* Special Projects Office, Bureau of Ordnance, Department of the Navy, Washington, July, 1958.
2. Malcolm, D. G., J. H. Roseboom, C. E. Clark, and W. Fazar, "Applications

of a Technique for R and D Program Evaluation," (PERT) *Operations Research*, Vol. VII, No. 5, September–October (1959) pp. 646-669.

3. Freund, J. E., and Williams, F. J. *Modern Business Statistics*, Pitman, 1959.

4. Miller, I., J. E. Freund, *Probability and Statistics for Engineers*, Prentice-Hall, Inc., 1965.

5. Moder, J. J., and E. G. Rodgers, "Judgment Estimates of the Moments of PERT Type Distributions," *Management Science*, Vol. 15, No. 2, October (1968).

6. Clark, Charles E., "The PERT Model for the Distribution of an Activity Time," *Operations Research*, Vol. 10, No. 3, May–June, 1962, pp. 405 and 406.

7. MacCrimmon, K. R., and C. A. Ryavec, "An Analytical Study of the PERT Assumptions," *Operations Research*, Vol. 12, No. 1, January–February 1964, pp. 16-37.

8. King, W. R., and T. A. Wilson, "Subjective Time Estimates in Critical Path Planning—A Preliminary Analysis," *Management Science*, Vol. 13, No. 5, January 1967, pp. 307-320.

9. Klingel, A. R., "Bias in PERT Project Completion Time Calculations for a Real Network," *Management Science*, Vol. 13, No. 4, December 1966, pp. B-194-201.

10. McMillan and Gonzalez. *Systems Analysis, A Computer Approach*, Irwin Publishing Co., 1965.

11. Clark, C. E., "The Greatest of a Finite Set of Random Variables," *Operations Research*, Vol. 9, No. 2, March–April (1961) pp. 145-162.

12. Moder, J. J. and C. R. Phillips, *Project Management with CPM and PERT*, Reinhold Corp., 1964, pp. 229-239.

13. Fulkerson, D. R., "Expected Critical Path Lengths in PERT Networks," *Operations Research*, Vol. 10, No. 6, November–December (1962), pp. 808–817.

14. Clingen, C. T., "A Modification of Fulkerson's Algorithm for Expected Duration of a PERT Project when Activities Have Continuous D.F.," *Operations Research*, Vol. 12, No. 4 (1964) pp. 629-632.

15. Elmaghraby, S. E., 'On the Expected Duration of PERT Type Networks," *Management Science*, Vol. 13, No. 5, January (1967) pp. 299-306.

16. Charnes, A., W. W. Cooper, and G. L. Thompson, Critical Path Analyses via Chance Constrained and Stochastic Programming," *Operations Research*, Vol. 12, No. 3, May–June (1964), pp. 460-70.

17. Martin, J. J., "Distribution of the Time Through a Directed, Acyclic Network," *Operations Research*, Vol. 13, No. 1, January–February (1965), pp. 46-66.

18. Hartley, H. O., and A. W. Wortham, "A Statistical Theory for PERT Critical Path Analysis," *Management Science*, Vol. 12, No. 10, June (1966), pp. B-469-481.

19. Van Slyke, R. M., "Monte Carlo Methods and the PERT Problem,"

Operations Research, Vol. 11, No. 5, September–October (1963), pp. 839-860.

EXERCISES

1. Verify the Central Limit Theorem by sampling, that is, by tossing three dice and recording the results on graphs, such as those shown in Figure 11-6. For convenience, use a white, a red, and a green die. Call the number of spots on the white die, X, and plot on the first figure. Call the number of spots on the white plus the red die, Y, and plot on the second figure. Finally, call the number of spots on all three dice, Z, and plot on the third figure. Compare the results of your experiment with the theoretical values given in Figure 11-6.

2. Verify the expected time and variance for event (4004-199), and verify the probability of 0.12 given for activity (4004-743)−(4004-199) in Figure 11-13b. Note: the values of a and b given in this figure are the end points of the distribution of activity performance time as used in conventional PERT Hence $(V_t)^{1/2} = (b - a)/6$ should be used in place of equation (3) given in this text. Also, note that the time interval 12–13–61 to 12-25-61 is equivalent to 1.6 working weeks.

3. Consider the oversimplified network given in Figure 11-19, which might be only a portion of a larger network, a portion which is subject to considerable chance variation in the performance times. In Figure 11-19, a and b are 5 and 95 percentiles, respectively.
   a. Compute t_e and V_t for each of the four activities.
   b. What is the earliest expected time of event 3?
   c. What is the variance, V_T, for the actual occurrence time for event 3?
   d. What is the probability that the project will be completed by time 8? By time 9?
   e. What time are you fairly sure (say 95 percent confident) of meeting for the completion of the project, i.e., the occurrence of event 3?

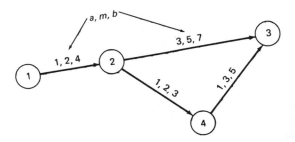

Figure 11-19

4. With reference to Figure 11-1, what is the probability that event 8 will be completed on or before the end of the 12th day, assuming all activities are started as early as possible? What scheduled time for the entire project would you feel 95 percent confident of meeting without having to expedite the project?

5. With reference to Figure 11-1, what is the (conditional) probability that event 8 will be completed on or before the end of the 16th day, assuming that a scheduled time of 10 is met on event 7?

6. Repeat exercise 5 on the assumption that event 7 has occurred 2 days late, i.e., at the end of the 12th day.

7. Solve exercise 3 using the Monte Carlo approach described in the text. For simplicity, assume that the distribution of performance times for each activity is rectangular between the limits given by a and b. Using a table of random numbers that are uniformly distributed on the range from 0 to 1, given in most standard statistics texts, transform them to the distribution required for each of the activities in Figure 11-19. For example, activity 2–3 has a possible range of 3 to 7, or 4 time units, and a mean of 5. Letting r denote a random number uniformly distributed on the interval 0 to 1, then, $t = 3 + 4r$, will be uniformly distributed on the range 3 to 7 as desired. Decide for yourself how many simulations of the network should be made, and express the results in a form such as given in Figures 11-17 and 18. Discuss your results.

8. Repeat exercise 7 for the network given in Figure 11-1 making the same assumptions as suggested in exercise 7 about the distributions of activity performance times. To carry out this exercise, use the following values of a, m, and b.

Activity	a	m	b
0–1	1.5	2	2.5
0–3	1	2	3
0–6	1	1	1
1–2	3	4	5
2–5	1	1	1
3–4	2	5	8
3–7	6	8	10
4–5	3	4	5
5–8	2	3	4
6–7	2	3	4
7–8	3.5	5	6.5

a. Applying the rule of thumb given in the text on merge event bias, will the latter be significant in this problem? Why? At what merge events?

b. If we wanted to simplify the Monte Carlo simulation, would the elimination of slack paths 0-6-7 and 0-1-2-5 be justified?

c. Perform the Monte Carlo simulation and comment on the results obtained.

9. Referring to exercise 3 in Chapter 9, assume the times given under a and b are 0 and 100 percentiles, so that the standard deviation, $(V_t)^{1/2} = (b-a)/6$, and the times under the m column are single time estimates of the mean, i.e., $t_e = m$, in this case. Without making a *complete* PERT statistical analysis, answer the following questions.

a. Is there a significant merge event bias problem for this network?

b. What is the mean and *approximate* standard deviation of the total time for this project?

c. What are the *approximate* chances of completing this project in 400 hours?

Appendix 11-1a
The Cumulative Normal Distribution Function†

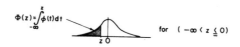

$$\Phi(z) = \int_{-\infty}^{z} \phi(t)\,dt \qquad \text{for } (-\infty < z \leq 0)$$

z	.00	.01	.02	.03	.04	.05	.06	.07	.08	.09
−.0	.5000	.4960	.4920	.4880	.4840	.4801	.4761	.4721	.4681	.4641
−.1	.4602	.4562	.4522	.4483	.4443	.4404	.4364	.4325	.4286	.4247
−.2	.4207	.4168	.4129	.4090	.4052	.4013	.3974	.3936	.3897	.3859
−.3	.3821	.3783	.3745	.3707	.3669	.3632	.3594	.3557	.3520	.3483
−.4	.3446	.3409	.3372	.3336	.3300	.3264	.3228	.3192	.3156	.3121
−.5	.3085	.3050	.3015	.2981	.2946	.2912	.2877	.2843	.2810	.2776
−.6	.2743	.2709	.2676	.2643	.2611	.2578	.2546	.2514	.2483	.2451
−.7	.2420	.2389	.2358	.2327	.2297	.2266	.2236	.2206	.2177	.2148
−.8	.2119	.2090	.2061	.2033	.2005	.1977	.1949	.1922	.1894	.1867
−.9	.1841	.1814	.1788	.1762	.1736	.1711	.1685	.1660	.1635	.1611
−1.0	.1587	.1562	.1539	.1515	.1492	.1469	.1446	.1423	.1401	.1379
−1.1	.1357	.1335	.1314	.1292	.1271	.1251	.1230	.1210	.1190	.1170
−1.2	.1151	.1131	.1112	.1093	.1075	.1056	.1038	.1020	.1003	.09853
−1.3	.09680	.09510	.09342	.09176	.09012	.08851	.08691	.08534	.08379	.08226
−1.4	.08076	.07927	.07780	.07636	.07493	.07353	.07215	.07078	.06944	.06811
−1.5	.06681	.06552	.06426	.06301	.06178	.06057	.05938	.05821	.05705	.05592
−1.6	.05480	.05370	.05262	.05155	.05050	.04947	.04846	.04746	.04648	.04551
−1.7	.04457	.04363	.04272	.04182	.04093	.04006	.03920	.03836	.03754	.03673
−1.8	.03593	.03515	.03438	.03362	.03288	.03216	.03144	.03074	.03005	.02938
−1.9	.02872	.02807	.02743	.02680	.02619	.02559	.02500	.02442	0.2385	.02330
−2.0	.02275	.02222	.02169	.02118	.02068	.02018	.01970	.01923	.01876	.01831
−2.1	.01786	.01743	.01700	.01659	.01618	.01578	.01539	.01500	.01463	.01426
−2.2	.01390	.01355	.01321	.01287	.01255	.01222	.01191	.01160	.01130	.01101
−2.3	.01072	.01044	.01017	$.0^2 9903$	$.0^2 9642$	$.0^2 9387$	$.0^2 9137$	$.0^2 8894$	$.0^2 8656$	$.0^2 8424$
−2.4	$.0^2 8198$	$.0^2 7976$	$.0^2 7760$	$.0^2 7549$	$.0^2 7344$	$.0^2 7143$	$.0^2 6947$	$.0^2 6756$	$.0^2 6569$	$.0^2 6387$
−2.5	$.0^2 6210$	$.0^2 6037$	$.0^2 5868$	$.0^2 5703$	$.0^2 5543$	$.0^2 5386$	$.0^2 5234$	$.0^2 5085$	$.0^2 4940$	$.0^2 4799$
−2.6	$.0^2 4661$	$.0^2 4527$	$.0^2 4396$	$.0^2 4269$	$.0^2 4145$	$.0^2 4025$	$.0^2 3907$	$.0^2 3793$	$.0^2 3681$	$.0^2 3573$
−2.7	$.0^2 3467$	$.0^2 3364$	$.0^2 3264$	$.0^2 3167$	$.0^2 3072$	$.0^2 2980$	$.0^2 2890$	$.0^2 2803$	$.0^2 2718$	$.0^2 2635$
−2.8	$.0^2 2555$	$.0^2 2477$	$.0^2 2401$	$.0^2 2327$	$.0^2 2256$	$.0^2 2186$	$.0^2 2118$	$.0^2 2052$	$.0^2 1988$	$.0^2 1926$
−2.9	$.0^2 1866$	$.0^2 1807$	$.0^2 1750$	$.0^2 1695$	$.0^2 1641$	$.0^2 1589$	$.0^2 1538$	$.0^2 1489$	$.0^2 1441$	$.0^2 1395$
−3.0	$.0^3 1350$	$.0^3 1306$	$.0^3 1264$	$.0^3 1223$	$.0^3 1183$	$.0^3 1144$	$.0^3 1107$	$.0^3 1070$	$.0^3 1035$	$.0^3 1001$
−3.1	$.0^3 9676$	$.0^3 9354$	$.0^3 9043$	$.0^3 8740$	$.0^3 8447$	$.0^3 8164$	$.0^3 7888$	$.0^3 7622$	$.0^3 7364$	$.0^3 7114$
−3.2	$.0^3 6871$	$.0^3 6637$	$.0^3 6410$	$.0^3 6190$	$.0^3 5976$	$.0^3 5770$	$.0^3 5571$	$.0^3 5377$	$.0^3 5190$	$.0^3 5009$
−3.3	$.0^3 4834$	$.0^3 4665$	$.0^3 4501$	$.0^3 4342$	$.0^3 4189$	$.0^3 4041$	$.0^3 3897$	$.0^3 3758$	$.0^3 3624$	$.0^3 3495$
−3.4	$.0^3 3369$	$.0^3 3248$	$.0^3 3131$	$.0^3 3018$	$.0^3 2909$	$.0^3 2803$	$.0^3 2701$	$.0^3 2602$	$.0^3 2507$	$.0^3 2415$
−3.5	$.0^3 2326$	$.0^3 2241$	$.0^3 2158$	$.0^3 2078$	$.0^3 2001$	$.0^3 1926$	$.0^3 1854$	$.0^3 1785$	$.0^3 1718$	$.0^3 1653$
−3.6	$.0^3 1591$	$.0^3 1531$	$.0^3 1473$	$.0^3 1417$	$.0^3 1363$	$.0^3 1311$	$.0^3 1261$	$.0^3 1213$	$.0^3 1166$	$.0^3 1121$
−3.7	$.0^4 1078$	$.0^4 1036$	$.0^4 9961$	$.0^4 9574$	$.0^4 9201$	$.0^4 8842$	$.0^4 8496$	$.0^4 8162$	$.0^4 7841$	$.0^4 7532$
−3.8	$.0^4 7235$	$.0^4 6948$	$.0^4 6673$	$.0^4 6407$	$.0^4 6152$	$.0^4 5906$	$.0^4 5669$	$.0^4 5442$	$.0^4 5223$	$.0^4 5012$
−3.9	$.0^4 4810$	$.0^4 4615$	$.0^4 4427$	$.0^4 4247$	$.0^4 4074$	$.0^4 3908$	$.0^4 3747$	$.0^4 2594$	$.0^4 3446$	$.0^4 3304$
−4.0	$.0^4 3167$	$.0^4 3036$	$.0^4 2910$	$.0^4 2789$	$.0^4 2673$	$.0^4 2561$	$.0^4 2454$	$.0^4 2351$	$.0^4 2252$	$.0^4 2157$
−4.1	$.0^4 2066$	$.0^4 1987$	$.0^4 1894$	$.0^4 1814$	$.0^4 1737$	$.0^4 1662$	$.0^4 1591$	$.0^4 1523$	$.0^4 1458$	$.0^4 1395$
−4.2	$.0^4 1335$	$.0^4 1277$	$.0^4 1222$	$.0^4 1168$	$.0^4 1118$	$.0^4 1069$	$.0^4 1022$	$.0^5 9774$	$.0^5 9345$	$.0^5 8934$
−4.3	$.0^5 8540$	$.0^5 8163$	$.0^5 7801$	$.0^5 7455$	$.0^5 7124$	$.0^5 6807$	$.0^5 6503$	$.0^5 6212$	$.0^5 5934$	$.0^5 5668$
−4.4	$.0^5 5413$	$.0^5 5169$	$.0^5 4935$	$.0^5 4712$	$.0^5 4498$	$.0^5 4294$	$.0^5 4098$	$.0^5 3911$	$.0^5 3732$	$.0^5 3561$
−4.5	$.0^5 3398$	$.0^5 3241$	$.0^5 3092$	$.0^5 2949$	$.0^5 2813$	$.0^5 2682$	$.0^5 2558$	$.0^5 2439$	$.0^5 2325$	$.0^5 2216$
−4.6	$.0^5 2112$	$.0^5 2013$	$.0^5 1919$	$.0^5 1828$	$.0^5 1742$	$.0^5 1660$	$.0^5 1581$	$.0^5 1506$	$.0^5 1434$	$.0^5 1366$
−4.7	$.0^5 1301$	$.0^5 1239$	$.0^5 1179$	$.0^5 1123$	$.0^5 1069$	$.0^5 1017$	$.0^6 9680$	$.0^6 9211$	$.0^6 8765$	$.0^6 8339$
−4.8	$.0^6 7933$	$.0^6 7547$	$.0^6 7178$	$.0^6 6827$	$.0^6 6492$	$.0^6 6173$	$.0^6 5869$	$.0^6 5580$	$.0^6 5304$	$.0^6 5042$
−4.9	$.0^6 4792$	$.0^6 4554$	$.0^6 4327$	$.0^6 4111$	$.0^6 3906$	$.0^6 3711$	$.0^6 3525$	$.0^6 3348$	$.0^6 3179$	$.0^6 3019$

Example: $\Phi(-3.57) = .0^3 1785 = 0.0001785.$

† By permission from A. Hald, *Statistical Tables, and Formulas,* John Wiley & Sons. Inc., New York, 1952.

Appendix 11-1b
The Cumulative Normal Distribution Function†

$$\Phi(z) = \int_{-\infty}^{z} \phi(t)\,dt \qquad \text{for } (0 \leq z < \infty)$$

z	.00	.01	.02	.03	.04	.05	.06	.07	.08	.09
.0	.5000	.5040	.5080	.5120	.5160	.5199	.5239	.5279	.5319	.5359
.1	.5398	.5438	.5478	.5517	.5557	.5596	.5636	.5675	.5714	.5753
.2	.5793	.5832	.5871	.5910	.5948	.5987	.6026	.6064	.6103	.6141
.3	.6179	.6217	.6255	.6293	.6331	.6368	.6406	.6443	.6480	.6517
.4	.6554	.6591	.6628	.6664	.6700	.6736	.6772	.6808	.6844	.6879
.5	.6915	.6950	.6985	.7019	.7054	.7088	.7123	.7157	.7190	.7224
.6	.7257	.7291	.7324	.7357	.7389	.7422	.7454	.7486	.7517	.7549
.7	.7580	.7611	.7642	.7673	.7703	.7734	.7764	.7794	.7823	.7852
.8	.7881	.7910	.7939	.7967	.7995	.8023	.8051	.8078	.8106	.8133
.9	.8159	.8186	.8212	.8238	.8264	.8289	.8315	.8340	.8365	.8389
1.0	.8413	.8438	.8461	.8485	.8508	.8531	.8554	.8577	.8599	.8621
1.1	.8643	.8665	.8686	.8708	.8729	.8749	.8770	.8790	.8810	.8830
1.2	.8849	.8869	.8888	.8907	.8925	.8944	.8962	.8980	.8997	.90147
1.3	.90320	.90490	.90658	.90824	.90988	.91149	.91309	.91466	.91621	.91774
1.4	.91924	.92073	.92220	.92364	.92507	.92647	.92785	.92922	.93056	.93189
1.5	.93319	.93448	.93574	.93699	.93822	.93943	.94062	.94179	.94295	.94408
1.6	.94520	.94630	.94738	.94845	.94950	.95053	.95154	.95254	.95352	.95449
1.7	.95543	.95637	.95728	.95818	.95907	.95994	.96080	.96164	.96246	.96327
1.8	.96407	.96485	.96562	.96638	.96712	.96784	.96856	.96926	.96995	.97062
1.9	.97128	.97193	.97257	.97320	.97381	.97441	.97500	.97558	.97615	.97670
2.0	.97725	.97778	.97831	.97882	.97932	.97982	.98030	.98077	.98124	.98169
2.1	.98214	.98257	.98300	.98341	.98382	.98422	.98461	.98500	.98537	.98574
2.2	.98610	.98645	.98679	.98713	.98745	.98778	.98809	.98840	.98870	.98899
2.3	.98928	.98956	.98983	$.9^2 0097$	$.9^2 0358$	$.9^2 0613$	$.9^2 0863$	$.9^2 1106$	$.9^2 1344$	$.9^2 1576$
2.4	$.9^2 1802$	$.9^2 2024$	$.9^2 2240$	$.9^2 2451$	$.9^2 2656$	$.9^2 2857$	$.9^2 3053$	$.9^2 3244$	$.9^2 3431$	$.9^2 3613$
2.5	$.9^2 3790$	$.9^2 3963$	$.9^2 4132$	$.9^2 4297$	$.9^2 4457$	$.9^2 4614$	$.9^2 4766$	$.9^2 4915$	$.9^2 5060$	$.9^2 5201$
2.6	$.9^2 5339$	$.9^2 5473$	$.9^2 5604$	$.9^2 5731$	$.9^2 5855$	$.9^2 5975$	$.9^2 6093$	$.9^2 6207$	$.9^2 6319$	$.9^2 6427$
2.7	$.9^2 6533$	$.9^2 6636$	$.9^2 6736$	$.9^2 6833$	$.9^2 6928$	$.9^2 7020$	$.9^2 7110$	$.9^2 7197$	$.9^2 7282$	$.9^2 7365$
2.8	$.9^2 7445$	$.9^2 7523$	$.9^2 7599$	$.9^2 7673$	$.9^2 7744$	$.9^2 7814$	$.9^2 7882$	$.9^2 7948$	$.9^2 8012$	$.9^2 8074$
2.9	$.9^2 8134$	$.9^2 8193$	$.9^2 8250$	$.9^2 8305$	$.9^2 8359$	$.9^2 8411$	$.9^2 8462$	$.9^2 8511$	$.9^2 8559$	$.9^2 8605$
3.0	$.9^2 8650$	$.9^2 8694$	$.9^2 8736$	$.9^2 8777$	$.9^2 8817$	$.9^2 8856$	$.9^2 8893$	$.9^2 8930$	$.9^2 8965$	$.9^2 8999$
3.1	$.9^3 0324$	$.9^3 0646$	$.9^3 0957$	$.9^3 1260$	$.9^3 1553$	$.9^3 1836$	$.9^3 2112$	$.9^3 2378$	$.9^3 2636$	$.9^3 2886$
3.2	$.9^3 3129$	$.9^3 3363$	$.9^3 3590$	$.9^3 3810$	$.9^3 4024$	$.9^3 4230$	$.9^3 4429$	$.9^3 4623$	$.9^3 4810$	$.9^3 4991$
3.3	$.9^3 5166$	$.9^3 5335$	$.9^3 5499$	$.9^3 5658$	$.9^3 5811$	$.9^3 5959$	$.9^3 6103$	$.9^3 6242$	$.9^3 6376$	$.9^3 6505$
3.4	$.9^3 6631$	$.9^3 6752$	$.9^3 6869$	$.9^3 6982$	$.9^3 7091$	$.9^3 7197$	$.9^3 7299$	$.9^3 7398$	$.9^3 7493$	$.9^3 7585$
3.5	$.9^3 7674$	$.9^3 7759$	$.9^3 7842$	$.9^3 7922$	$.9^3 7999$	$.9^3 8074$	$.9^3 8146$	$.9^3 8215$	$.9^3 8282$	$.9^3 8347$
3.6	$.9^3 8409$	$.9^3 8469$	$.9^3 8527$	$.9^3 8583$	$.9^3 8637$	$.9^3 8689$	$.9^3 8739$	$.9^3 8787$	$.9^3 8834$	$.9^3 8879$
3.7	$.9^3 8922$	$.9^3 8964$	$.9^4 0039$	$.9^4 0426$	$.9^4 0799$	$.9^4 1158$	$.9^4 1504$	$.9^4 1838$	$.9^4 2159$	$.9^4 2468$
3.8	$.9^4 2765$	$.9^4 3052$	$.9^4 3327$	$.9^4 3593$	$.9^4 3848$	$.9^4 4094$	$.9^4 4331$	$.9^4 4558$	$.9^4 4777$	$.9^4 4988$
3.9	$.9^4 5190$	$.9^4 5385$	$.9^4 5573$	$.9^4 5753$	$.9^4 5926$	$.9^4 6092$	$.9^4 6253$	$.9^4 6406$	$.9^4 6554$	$.9^4 6696$
4.0	$.9^4 6833$	$.9^4 6964$	$.9^4 7090$	$.9^4 7211$	$.9^4 7327$	$.9^4 7439$	$.9^4 7546$	$.9^4 7649$	$.9^4 7748$	$.9^4 7843$
4.1	$.9^4 7934$	$.9^4 8022$	$.9^4 8106$	$.9^4 8186$	$.9^4 8263$	$.9^4 8338$	$.9^4 8409$	$.9^4 8477$	$.9^4 8542$	$.9^4 8605$
4.2	$.9^4 8665$	$.9^4 8723$	$.9^4 8778$	$.9^4 8832$	$.9^4 8882$	$.9^4 8931$	$.9^4 8978$	$.9^5 0226$	$.9^5 0655$	$.9^5 1066$
4.3	$.9^5 1460$	$.9^5 1837$	$.9^5 2199$	$.9^5 2545$	$.9^5 2876$	$.9^5 3193$	$.9^5 3497$	$.9^5 3788$	$.9^5 4066$	$.9^5 4332$
4.4	$.9^5 4587$	$.9^5 4831$	$.9^5 5065$	$.9^5 5288$	$.9^5 5502$	$.9^5 5706$	$.9^5 5902$	$.9^5 6089$	$.9^5 6268$	$.9^5 6439$
4.5	$.9^5 6602$	$.9^5 6759$	$.9^5 6908$	$.9^5 7051$	$.9^5 7187$	$.9^5 7318$	$.9^5 7442$	$.9^5 7561$	$.9^5 7675$	$.9^5 7784$
4.6	$.9^5 7888$	$.9^5 7987$	$.9^5 8081$	$.9^5 8172$	$.9^5 8258$	$.9^5 8340$	$.9^5 8419$	$.9^5 8494$	$.9^5 8566$	$.9^5 8634$
4.7	$.9^5 8699$	$.9^5 8761$	$.9^5 8821$	$.9^5 8877$	$.9^5 8931$	$.9^5 8983$	$.9^6 0320$	$.9^6 0789$	$.9^6 1235$	$.9^6 1661$
4.8	$.9^6 2067$	$.9^6 2453$	$.9^6 2822$	$.9^6 3173$	$.9^6 3508$	$.9^6 3827$	$.9^6 4131$	$.9^6 4420$	$.9^6 4696$	$.9^6 4958$
4.9	$.9^6 5208$	$.9^6 5446$	$.9^6 5673$	$.9^6 5889$	$.9^6 6094$	$.9^6 6289$	$.9^6 6475$	$.9^6 6652$	$.9^6 6821$	$.9^6 6981$

Example: $\Phi(3.57) = .9^3 8215 = 0.9998215.$

† By permission from A. Hald, *Statistical Tables, and Formulas,* John Wiley & Sons, Inc., New York, 1952.

APPENDIX 11-2
USE OF
HISTORICAL
DATA IN
ESTIMATING
a, m, AND b

Occasionally one may have historical (sample) activity duration data on which to base estimates of t_e and $(V_t)^{1/2}$, or better, to estimate a, m, and b, which when processed in the usual manner, with equations (3) and (4), will give the desired estimates of $(V_t)^{1/2}$ and t_e. This procedure has merit, if the following conditions are satisfied.

(1) The historical data are representative of the hypothetical population (Figures 11-1 and 11-8) to be "sampled" in the future for the activity in question; that is, the activity is precisely the same, and the conditions which prevailed during the collection of the historical data are representative of those expected to prevail in

the future when the activity in question is to be performed.
(2) The sample of historical data is of "sufficient" size. Quantitative specification of what is "sufficient" depends on the nature of the activity in question and the experience and abilities of the person supplying the estimates; however, a sample of less than four or five observations would generally not be considered "sufficient."

If the above assumptions are satisfied, estimates of a, m, and b can be obtained from-equations (7), (8), and (9) below, wherein

R = range of sample data
   = largest observation − smallest observation
k = $3/d_2$, where d_2 is the statistical quality control constant tabled as a function of the number, n, of activity times in the sample data. Actually, d_2 is the average of the ratio $R/(V_t)^{1/2}$. Values of k are given in Table 11-7, and are used to compute the constants a and b.
$\bar{t}$ = arithmetic average of the sample data

$$\text{Estimate of } m = \bar{t} \tag{7}$$

$$\text{Estimate of } a = \bar{t} - kR \tag{8}$$

$$\text{Estimate of } b = \bar{t} + kR \tag{9}$$

Table 11-7. Constant to Convert the Range to Estimates of the Standard Deviation

Sample Size†	(Range/Std. Dev.) = d_2††	$k = 1.6/d_2$
2	1.13	1.416
3	1.69	0.947
4	2.06	0.777
5	2.33	0.687
6	2.53	0.632
7	2.70	0.593
8	2.85	0.561
9	2.97	0.539
10	3.08	0.519
12	3.26	0.491
15	3.47	0.461
20	3.74	0.428
25	3.93	0.407

† Although this table includes samples as small as two, one should not rely solely on the sample data unless the sample size is at least four.
†† The symbol d_2 used here is the universal designation of this ratio, which is widely used and tabled in statistical quality control literature; it assumes the random variable is normally distributed.

In situations where kR is greater than $\overline{t}$, and hence a as given by equation (8) is negative, it is suggested that the following be used.

$$a = 0 \qquad\qquad (8a)$$

$$b = 2\overline{t} \qquad\qquad (9a)$$

12

SUMMARY
COMMENTS ON
PRACTICAL
APPLICATIONS

One might say that critical path methods are applicable to the management of a project from the cradle to the grave. At each of the stages in the life of a project, there are a number of areas related to the practical applications of critical path methods which are not covered in the previous chapters but are deserving of attention. This final chapter will treat these points under the headings of Use of Critical Path Methods in Preparing Project Proposals, Implementation, Project Control, and Multiproject Scheduling. In conclusion, the authors summarize the objectives of this text and hazard some forecasts of the future in project management techniques.

USE OF CRITICAL PATH METHODS IN
PREPARING PROJECT PROPOSALS

The numerous potential applications of critical path methods in the preparation of project proposals are fairly obvious and straightforward. These applications are based on the various project planning and scheduling techniques discussed in this text, except perhaps that these techniques are applied with less detail than would normally be used on a project that is definitely to be executed. If the project is to be executed, the proposal network serves as a framework for developing a detailed plan and schedule to be used in carrying out the project.

In cases where the project completion time is specified in the contract, critical path methods are useful in determining a project plan which will meet the time specification. It may turn out that this plan requires the use of certain time saving features which add unexpected costs, or risks, to those which would normally be required to complete the project tasks. For example, if performace time is extremely critical, "crash" time performance of critical path activities may be required. In addition, it may be necessary to perform certain activities concurrently, which would normally be performed in series. An example of this is shown in Figure 12-1, in which a "normal" plan and an expedited plan are shown for the same project. The added risks of the expedited plan over the normal plan must, of course, be considered in preparing the proposal cost estimate.

If the performance times for the activities which make up the project are subject to a considerable amount of random variation, then the PERT statistical approach is appropriate. As described in Chapter 11, the basic project plan can be developed, using single time estimates of the activity performance times. Then, by obtaining optimistic and pessimistic time estimates for the activities on the critical path, the approximate probability of completing the project (or subprojects) on schedule can be computed. If this probability is low, even when an expedited plan such as shown in Figure 12-1 is used, then entering into such a contract would indeed be risky. If it is also known that contract specifications and time requirements are reasonable, then the project plan does not embody the necessary "technology" to satisfactorily pursue this contract. However, if the probability of meeting the scheduled times is high using the expedited project plan, but the probability is low using the normal plan, then one can interpret the latter as indicating the chances that some or all of the extra expenses involved in the expedited plan will be required to complete the project on schedule. The ideal

Normal plan

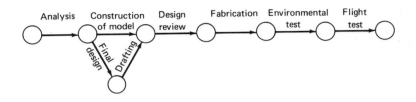

Expedited plan

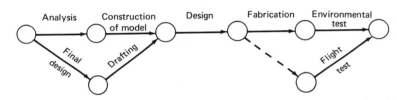

Figure 12-1 Normal plan and expedited plan for a hypothetical rocket project.

situation would, of course, be a high probability of meeting the schedules using a "normal" project plan.

If the proposal requires consideration of alternate completion times and costs, then the time-cost trade-off procedures described in Chapter 9 may be appropriate in arriving at project costs. Similarly, if the project under consideration will be competing with a fixed set of resources, then the procedures discussed under multiproject control (later in this chapter) are applicable. It may turn out that the proposed project schedule dovetails nicely with the phasing out of current projects, or it may put an extreme burden on certain critical resources. This type of analysis may greatly influence the profit margin that management places on the project, or influence the decision of whether or not to submit a bid on the proposed project.

A number of firms in various industries are now using critical path methods routinely, in varying degrees of detail, in preparing project proposals. One large metal-working firm requires that all internal proposals for capital expenditures over $25,000 be accompanied by a CPM network showing the project plan and schedule. Some advertising firms submit to prospective clients networks of proposed promotional campaigns, showing how the activities involved in product distribution, space advertising, television and radio commercials, surveys, and other facets in the campaign are to be coordinated. The implementation of a computerized management system has proved, in many organizations, to be a type of project that is difficult to schedule and control; now it is not unusual for project networks to be submitted to management, along with flow charts, proposed report formats, and other elements of proposed projects, and for the networks to be updated weekly or bi-weekly to maintain status control.

CONTRACTUAL REQUIREMENTS

As mentioned in Chapter 1 and elsewhere, the use of network methods has been boosted greatly by contractual requirements that CPM or PERT be used to report the plans and progress of projects. This has been particularly true in the two largest project-oriented industries, aerospace and construction. In the aerospace industry, where the primary customer is the U. S. Government and its various military and space agencies, the accommodation to network requirements was widespread and relatively quick. Contractors in the aerospace industry are accustomed to extensive Government reporting requirements: the firms are relatively large and equipped with computers and capable technical staffs; and the contracts are often on a cost-plus basis. For these reasons, contractors were able to meet the first network requirements almost immediately. Some projects produced networks of several thousand activities in a few months.

In the construction industry, however, the circumstances have been quite different. The Government has played a major role here also, through the Corps of Engineers, the Bureau of Yards and Docks, NASA, and other agencies responsible for large construction programs. But construction companies tend to be relatively small, low-overhead organizations without computer equipment, or data processing personnel. Also, the contracts are usually on a fixed-price basis. Under these circumstances, which exist throughout the commercial as well as Government segments of the industry, requirements for critical path methods encounter a variety of difficulties.

The first problem in this application is to prepare specifications for the network reporting system desired. Since the contractors are often not familiar with CPM or PERT, are not administratively geared to handle increases in technical reporting, and often must pay for outside assistance, the firms will take very conservative approaches to critical path method specifications. The authors have seen specifications consisting of only one or two sentences, saying only that "the contractor shall report progress on the project monthly by means of a Critical Path Method chart and schedule." Requirements that are this brief, of course, invite the contractor to submit a five-activity network or anything that he thinks might meet the minimum technical requirements of a CPM "chart and schedule."

At the other extreme, one construction office of a Government agency prepared and repeatedly used standard CPM specifications that covered eight pages and contained almost 1000 words. This specification attempted to detail exactly how all the initial updating computations were to be made and how the computer reports were to be formatted (including the column headings for such items as an "actual early date" and other unique descriptions). These specifications were so confusing that only a few specialized computer service bureaus in the region could satisfy them, and the contractors were dependent upon the use of these service bureaus.

Although it is not practical to write general specifications that would apply to all types of construction projects, some guidelines can be offered to those who wish to prepare their own requirements. The considerations listed below, and the sample specifications given in Appendix 12-1, are suggested as guides to the preparation of reasonable and adequate requirements:

(1) *Definition of CPM.* An available document or textbook should be referenced as comprising a definition of the basic technique to be employed. The terms used throughout the specification should be consistent with the terms used in the reference.

(2) *Level of Detail.* The best way to specify the desired level of network detail is to specify a range of the number of activities to be included. The range selected, however, should be based on practical experience with the type of construction involved.

(3) *Output Reports.* Since critical path methods and computer programs vary in the specific data included in reports, the data desired should be spelled out, item for item. The desired sequences (sorts) of the reports should also be specified, taking care not to require more reports than will actually be used. Normally two sorts, by *I-J* and total slack, are adequate, and three should be a maximum.

(4) *Updating.* As described in Chapter 5, some updating procedures give misleading or erroneous results. Therefore, it may be necessary to reference or to fully describe the updating calculation required. The frequency of updating reports, number of copies desired, whether revised networks are required, and similar

details should be mentioned. In this connection, it should be noted that a requirement to maintain a network on a time scale will greatly increase the updating cost and may adversely affect the timeliness of the reports.

(5) *Cost Reporting.* If the network is also to be used for reporting the cost of work completed, the specific means of allocating all costs to activities and of determining the percent completion on activities must be fully detailed.

The sample specifications in Appendix 12-1 illustrate how most of the above elements may be handled, although the cost reporting aspect is not included.

IMPLEMENTATION

Generally speaking, the implementation of critical path methods involves the five steps outlined in Chapter 1 and summarized as follows: (1) planning, (2) initial scheduling, (3) analyzing resource utilization and time constraints, (4) final scheduling, and (5) controlling.

However, it should be emphasized that the full application of all the techniques presented in this text is not required in order to accomplish these five steps and obtain worthwhile improvements in the plan and conduct of projects. *Nor is the size of the project a critical factor in the practical economic implementation of critical path methods.* Networks sketched on the backs of envelopes have proved to be useful ways of quickly analyzing and communicating the plan of a small project, such as the preparation of a technical report. The use of additional techniques, such as the PERT statistical approach and time-cost trade-offs may also be applied in concept, if not in full detail, to relatively small projects. Thus, the implementation of critical path concepts can be considered a matter of routine management practice, rather than an investment justified only for large or complex projects.

The implementation of network planning and control techniques should take place after a preliminary study has been made to determine how the project tasks are to be broken down and assigned to key personnel within the organization or to subcontractors. In large projects, particularly in the aerospace industries, this preliminary study is quite important; it first requires that the overall mission and performance goals of the system be refined to the satisfaction of the systems engineer. Then, the functional analysis of the system can begin, which will lead to the design requirements for the proposed system configuration. The establishment of the base-line design requirements is a major milestone in the systems engineering portion of a program definition study. It is at this time that the formal application of PERT can be made most effectively.

Utilizing Personnel Effectively

The preparation of the project network is a job for the key management of the project, the person or team of persons who know the most about the objectives, technology, and resources of the project. To conserve the valuable time of these personnel, the networking effort should be a concerted and concentrated one, not a secondary activity that becomes drawn out and perhaps never completed. There is a period of time that management must devote to planning, and the network should be used as a vehicle for, an aid to, and a documentation of this valuable effort. Technicians can relieve much of the load on the project management by transcribing sketches into legible networks, by making computations, and by making preliminary analyses of the schedule. For this reason, many organizations have trained young men and women as part-time or full-time critical path analysts.

Working with Subcontractors

When subcontractors play a major role in a project, they should play a major role in the critical path planning, scheduling, and control. In some cases this may require group meetings to develop the network and time estimates. However, in many instances it is not practical to call in a group of people unfamiliar with network theory and expect them all to contribute effectively to the early draft of a network. In these circumstances it may be better for the prime contractor to develop the rough draft as far as possible, then call in the subcontractors to comment and add time estimates to their particular areas. It is often possible for a person to read and effectively criticize a network, even though he may not have the training or experience to fully develop a network. One may also elect to hold short courses in network preparation, then ask each subcontractor to develop a subnetwork of his portion of the project. The feasibility of this approach depends not only on the size of the project and the scope of interest of the subcontractors, but also on how complex are the interrelationships among the various areas of responsibility.

Incidentally, one of the important side effects of critical path applications is the fact that it brings the subcontractors and the prime contractor together to meet and discuss the project. The group generally discovers technical problems and begins to work in advance toward solutions of these, as well as in cooperation on the planning and scheduling of the project.

PROJECT CONTROL

Once a project is underway, the critical path network and schedule should serve as a guide to the accomplishment of each activity in proper sequence and on schedule. It is in the fundamental nature of projects, however, that activities will seldom start or finish exactly as scheduled. Therefore, updating the plan and schedule is an important link in the critical path concept. There are no rigorous or standard updating procedures in general use, except where computer programs are involved. The procedures offered here are suggestions regarding the general approach, with some specific recommendations for particular circumstances.

There are four functions performed in the process of updating the schedule alone (without regard to cost revision). These are:

(1) denoting actual progress on the network,
(2) revising the network logic or time estimates of uncompleted activities,
(3) recomputing the basic schedule of earliest and latest allowable times, and
(4) revising the scheduled activity start times and denoting new critical paths.

Item 2 simply requires the erasure of previous notation and replacement with the new. The other changes may be categorized as Progress Notation and Revised Scheduling.

Progress Notation

A field supervisor or other person close to the actual progress of a project should be assigned the responsibility of making progress notations directly on a copy of the network. This is the most reliable way to maintain accurate records for input to the network updating procedure. Requiring the responsible first-line supervisor to make these routine daily or weekly notations also helps insure that the network serves its purpose as a detailed schedule of work.

Exactly how the notations are made is not particularly important, as long as they are clearly understood and complete. To be complete it is necessary only that both the actual start and the actual finish of each activity be recorded. Percent completion notes are not normally required, except where cost control data are involved or where precedence diagramming rules are used.

One of the most common ways of denoting actual dates on a network is by means of slashed lines through the activities with date flags attached. An example is shown in Figure 4-8 of Chapter 4.

On the working copy of the new network used to make these notations, it is often useful to use colored pencils for the notation, using a different color for each updating period. Then when converting the notes to computer input forms, the person doing the encoding knows to pick up only the green dates, or whatever was the color of the most recent period.

If the node scheme is used, the actual start and finish dates may be marked on the left and right sides of the nodes, respectively. If the network is computed manually, the node symbol may be designed to allow spaces for the actual dates to be recorded.

Revised Scheduling

Hand computation of a revised and updated network requires no special arithmetic. One merely begins the forward pass computation at the last uncompleted activity on each path. However, it is important that the first dates computed for each path be consistent with the effective date of the computation, as explained in Chapter 5. That is, no ES or EF date can be earlier than the effective or "cut-off" date of the progress report. If the first uncompleted activity has not started (or finished), then its ES (or EF) date must be recorded as at least equal to the effective date, and the forward pass is carried on from there.

Hand computation does, however, require some attention to the mechanics of erasure, lest this humble function become a major problem. Erasure of all remaining schedule times is required, of course, when a complete recomputation is to be made. One technique that reduces the erasure problem is to make the scheduling computations on a reproducible print of the network, instead of on the original tracing. Erasures may be made on the reproducible print without damage to the arrows and nodes of the network. A modification of this technique calls for the discarding of the old reproducible print instead of erasing its schedule figures. This procedure is diagrammed in Figure 12-2.

Recomputation may be necessary every time a network is updated and/or revised. If the actual progress and expected future progress are very close to the network schedule, there is no need to make a completely revised computation. The critical path network protects itself, in a sense, from the need for many recomputations, even when significant

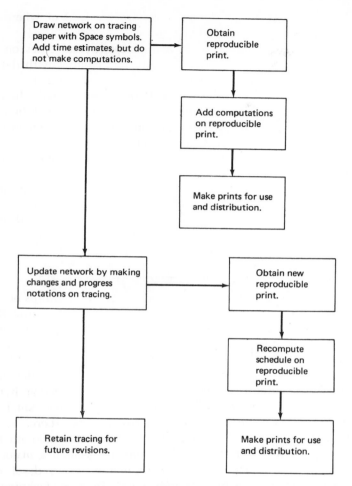

Figure 12-2   Flow diagram of a procedure for making revised hand computations.

delays or accelerations occur, for the network reveals most of the effects of schedule changes without recomputation. One need only to refer to the slack figures, for example, to see whether a specific delay will affect other portions of the project. With this information plainly visible, it is not always necessary to erase and recompute.

If the scheduling computations are made by computer, then the revision procedures specified by the computer program must be used. One of the problems common to most of the computer revision procedures is that indirect changes are required when certain other changes

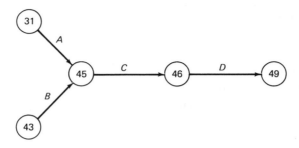

Figure 12-3a

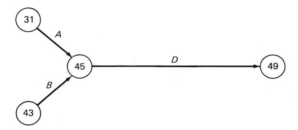

Figure 12-3b

are made. For example, consider Figure 12-3a. Assume that is desired to simply eliminate activity *C* from the network. Usually the program specifications will provide a "delete" code for this transaction. However, one must also remember to delete activity *D* and then add it in again with its new predecessor event. Therefore, deleting activity *C* in a typical program, as shown in Figure 12-3b, would require three revision cards (Table 12-1). Other revision procedures, such as changes in time estimates, are handled differently by the various computer programs and present no general problem.

Table 12-1. Revision Cards for the Deletion of Activity C

Transaction Code	Activity No.	Activity Description
Delete	45–46	activity *C*
Delete	46–49	activity *D*
Add	45–49	activity *D*

Calendar Dates

In most cases, practical use of critical path schedules require some translation between the times in working days and calendar days. When certain computer programs are used, the computer is given a project start (calendar) date, and the resulting schedule times are printed in both working days and calendar dates. When computations are made by hand, the usual procedure is to prepare a working day calendar for reference with the network schedule. The calendar is prepared by starting with the project start date and numbering the working days through the period of the project. Table 12-2 illustrates a portion of calendar for a project that began on Monday, January 6, 1969. In this calendar, the working day is assumed to be completed at the end of the corresponding calendar date.

Table 12-2. Portion of a Working Day Calendar for a Project that Began on January 6, 1969

Working Day	Calendar Days	Calendar Date
1	1	Jan. 6, 1969
2	2	7
3	3	8
4	4	9
5	5	10
—	6	11
—	7	12
6	8	13
7	9	14
8	10	15
•	•	•
•	•	•
•	•	•

Time-Scaled Networks

A "time-scaled" network is one in which the arrows and nodes are located by a time scale along the horizontal border. With the time scale, the length of each activity arrow (or its shadow projection on the horizontal scale) represents the activity's estimated duration. The location of the arrow represents its scheduled start and finish times. Similarly, the location of each node represents its scheduled occurrence time. The arrows and nodes may be located by their earliest times, latest times, or at selected times between these. Slack is customarily indicated by dashed lines.

The obvious advantage of time scaling is the visual clarity it provides for the analysis of concurrent activities. Time-scaled networks have revealed problems of concurrency (such as the intended use of a test facility by two groups at the same time) that were overlooked in the analysis of the schedule data in the tabulated form of a computer output. Concurrency problems may also be overlooked, where the schedules are denoted directly on the network in Space symbols, although there is less chance of error when the schedule and network are presented together in this manner. Time-scaled and condensed networks are also advantageous in presentations to top levels of management, since the schedule is communicated more quickly and more emphatically by the graphical technique.

The disadvantage of time-scaling is not so obvious but is nonetheless critical. The maintenance of time-scaled networks through updating and revision periods is expensive. A change in one time estimate or a delay in the actual progress on one path can change the location of a number of succeeding arrows and nodes, necessitating the redrawing of a large portion of the network. Under contracts that require frequent network revisions, the use of time-scaled networks could keep a technician or draftsman busy almost full time, and the networks could be delayed days or weeks in reaching the contracting agency.

Many attempts have been made to simplify network revision, so that time-scaled networks could be quickly and economically updated. Listed below are some of the materials and techniques that have been tried.

(1) *Steel panels with magnets for events, and tape or chalk lines for arrows.*   Networks constructed in this way have worked for some applications but cannot be easily reproduced, even by photography, for distribution.

(2) *Gummed labels and tape* which can be moved about on tracing paper. Networks made with these materials have tended to blister, peel, and result in poor reproductions.

(3) *Plastic panels and grease pencils.*   Again, the problem is reproduction for distribution.

(4) *Computer-drawn networks.*   The programming complexity involved has been the reason for only limited progress on this approach.

(5) *Mechanical Devices.*   Peg boards with wood or plastic strips for activities and rubber bands for interconnections have been used successfully for small networks(usually less than 50 activities).

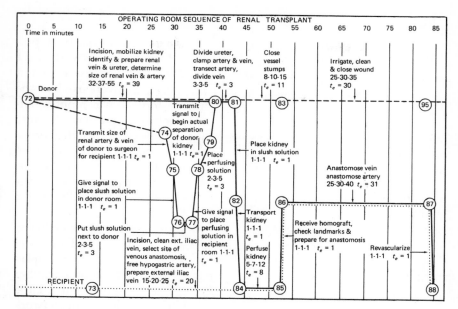

Figure 12-4 Time-scaled network of a kidney transplant procedure[1].

A commercial device called the Planalog system, consists of a metal board with horizontal channels in which the activities are placed. Vertical "fences" are incorporated to simulate the dependency relationships among the activities to restrict their horizontal movement between their early start and late finish times. An interesting application of this advice is described by H. Walker.[4]

Time-scaling can be particularly worthwhile where it is done as an aid to planning and not intended for use after the project is underway. An interesting example is shown in Figure 12-4, which is a network for a complex surgical procedure.[1] The only purpose of this network application is to help develop an efficient, well-coordinated plan, especially in the critical period between the 30th and 45th minutes. Once the plan is developed, the network has no further use.

Another example is shown in Figure 12-5.[2] Here three key resources in the project are summarized by time period, illustrating the utility of time-scaling networks where there are resource allocation problems. As in the previous example, this effort is usually worthwhile only in the planning stage of the project. In some cases, though, the resource problems are important enough to the time and cost economics of the project

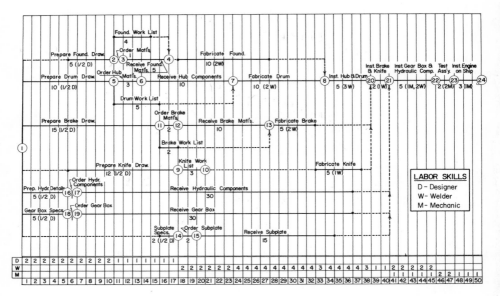

Figure 12-5 Time-scaled network with resource summary. Activities are scheduled to limit designer requirements to two and welders to four.[2]

that the time-scaled network should also be used to monitor the project progress. As mentioned above, however, project monitoring with time-scaled networks is practical, only if there are relatively few activities and the project management is centralized enough that copies of the network do not have to be distributed. It is expected that the development of computer hardware and software, which is within the state of the art today, will popularize the use of time-scaled networks. It is possible to display such networks on cathode ray tubes or draw them on paper with computer-controlled drafting machines.

MULTIPROJECT SCHEDULING

Almost any project may be said to consist of several subprojects and therefore, at least conceptually, there is no difference between project scheduling and multiproject scheduling. There are practical circumstances, however, in which multiproject scheduling has a significant meaning. One such circumstance is the common use of certain resources by projects that are otherwise independent. For example, two rocket research projects may be completely independent, except that they will

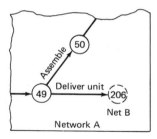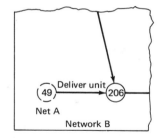

Figure 12-6   Example of interface notation.

both use the same test cell, or a road contractor may have several different jobs underway that will use the same paving machine. In these cases each project is scheduled separately, except for the activities dependent on the common resource. Other cases in which the term "multiproject scheduling" is applied are those in which two or more projects are part of the same over-all project objective but are separated by management responsibility or by the lack of capacity of the critical path computer program to process them as one network.

The approach to multiproject scheduling computations is basically simple. Events that are common to two or more networks are called "interfaces" and are denoted on each network by modified event symbols. The example in Figure 12-6 uses a dashed circle for interfaces. The forward pass and backward pass computations, if made by hand, are executed just as though the interfacing networks were completely integrated parts of the same network.

In some cases, however, the interfacing networks are not all available at the same time, and computations of one network must be made without the others. In the large scale research and development programs of aerospace industry, all the networks involved in a program are not developed simultaneously; some invariably are developed and processed months ahead of others. In these cases one may estimate (or obtain estimates from the groups responsible for the other networks) the earliest or latest occurrence time of the interfaces. For incoming interfaces, such as 49 in Network B, Figure 12-6, an estimate is needed of the *earliest expected start time* of activity 49-206, in order to process Network B. For outgoing interfaces, such as 206 in Network A, an estimate is needed of the *latest allowable finish time* of activity 49-206, in order to process Network A. If all estimates of this type are not available when the first of the two networks is to be processed, certain of the computations may be made anyway, and although the results will be incomplete, they will provide useful information. For example, if Network B were processed

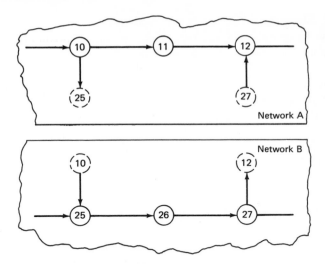

Figure 12-7 Example of changes in direction among interfaces in two networks.

without the earliest start time for activity 49-206, one could nonetheless compute the latest start time for this activity (assuming Network *B* has a final scheduled date from which to compute the backward pass). When Network *A* is developed, this latest time could then be used to completely process it, which would in turn supply the earliest start time for activity 49-206 and thus enable Network *B* to be completely processed.

Table 12-3.   Results of Iterative Computations on the Interfaces in Figure 12-7

| | | *Event Times Computed* | |
| | | Earliest $= E$ | Latest $= L$ |
Iteration	*Network*		
1	A	10, 11	none
2	B	25, 26, 27	none
3	A	12	12, 11
4	B	none	27, 26, 25
5	A	none	10

The use of this iterative procedure may be necessary when the capacity of a computer program will not permit the integrated processing of two networks. The procedure can become time-consuming and inconvenient if the number of interfaces, or rather the number of *changes in direction* of interfaces, becomes more than a few. To illustrate, consider the interface relationships indicated between the two networks in

Figure 12-7. Since there is one *change* in direction, five iterations would be required to completely process these networks separately, as indicated in Table 12-3 below, where event times are used for simplicity.

Clearly, even with only a few changes of interface direction, the large number of iterations required causes this approach to become prohibitively expensivee. Short cuts, such as partial hand computation or estimation of certain E and L values, are necessary in practice.

SUMMARY REMARKS

Objectives of the Text

In this textbook, the authors have attempted to present a broad range of topics in critical path methods and the role of these methods in the development of a discipline of project management. The historical development of the Critical Path Method and the Program Evaluation and Review Technique is explained. Although emphasis is placed on both the common aspects and the significant differences of these methods, the text material treats both CPM and PERT as historical variants on a "critical path methodology," which has asserted itself more clearly as the recent versions of the two original methods have become increasingly alike.

The basic concepts of critical path methodology are first presented in a manner intended to be understandable and immediately useful to project managers and supervisors at many levels and in diverse industries. In addition, an attempt has been made to provide a comprehensive treatment of the more complex aspects of critical path methodology, primarily directed to serious students of the subject, who may wish to design project management systems for particular applications. Included in these presentations are descriptions of certain existing systems.

The authors wish to emphasize, before closing this final chapter, that PERT and CPM network methods will not be effective if planning is incomplete or haphazard. The authors have observed initial applications of these techniques where they were forced upon the organization by contractual requirements. The half-hearted use of network methods in such cases are usually frustrating and unrewarding experiences. Networks produced under such conditions will very likely have false critical paths and will be difficult or impossible to monitor. Without a sincere effort to learn and get something out of the exercise, the use of network planning methods are not to be recommended.

Problems of implementing critical path methods are both psychological as well as technical, and the cost has been estimated to range from 0.2 to 1.0 percent of the total program cost. In a construction project the cost could be less than 0.2 percent. It has been estimated, however, that for most projects, critical path methods cost twice as much as the conventional planning methods that it replaces, and hence it must justify this increased cost. We feel the advantages of critical path methods, which have been discussed at the end of Chapter 1 as including better planning, communication, control, and other benefits, far outweigh this additional cost.

New ideas and techniques in this field are propagating rapidly. Undoubtedly, some significant advances will occur before this book is published. Some of the trends and anticipated advancements are forecasted below.

Future Developments in Critical Path Methods

Networking seems to be a natural mode of expressing project plans, and it is well on the way to becoming a standard part of the manager's language. In view of this trend, whenever the opportunity arises to change the accounting and planning systems of an organization—especially in project-oriented firms—the network approach should be built in. Network-based systems of cost and time estimating, accounting, and performance incentives, as well as project planning, scheduling and control, should be considered and exploited in the design of new management systems.

PERT has been used as a basis for incentive contracts by defining certain milestone type events as Incentive PERT Events.[3] These events are scored on the basis of time of occurrence and in some cases on performance as well. The percentage of these events accepted then determines the incentive bonus. Thus, a project with a normal fee of say 7 percent might vary from 0 to 14 percent according to performance.

In the construction industry the applications of CPM will certainly continue to expand. Eventually the industry will adopt certain standards for CPM specifications, which will help solve the problems of inconsistent applications of the method.

Although some development work continues to be done in the areas of time-cost trade-off and PERT statistics in order to improve the practical utility of these features, there is much more interest in the potential of network solutions to resource allocation problems. Some of the recent development work in resource allocation is still locked in the esoteric boxes of doctoral dissertations. However, as these ideas are brought out in the form of practical computer programs, the power of the network approach should be greatly enhanced. Already some computer service bureaus are marketing resource allocation services. In the near future we should see major improvements in computer software that will have capabilities for cost control, resource allocation, and a high degree of flexibility in formats for input and output. Such general purpose programs should open many new doors for critical path applications in the Seventies. These control systems are expected to incorporate the cathode ray tube as an input-output device more and more, as we approach more-or-less total management information systems. The latter will permit management to creatively interrogate the computer and carry out rather sophisticated analysis on a real time basis. This appears to hold real promise to the solution of the many communications problems of the functionally organized business that is heavily engaged in "project" type activities.

REFERENCES

1. Long, J. M., "Applying PERT/CPM to Complex Medical Procedures," *Proceedings of Seminar on Scientific Program Management*, Department of Industrial Engineering, Texas A & M University, June 1967.
2. Moder, Joseph J., "Applications of PERT and CPM in Multiproject Scheduling," *Proceedings of Seminar on Scientific Program Management*, Department of Industrial Engineering, Texas A & M University, June 1967.
3. Miller, Robert W., *Schedule, Cost, and Profit Control with PERT*, McGraw-Hill Book Co., New York, 1963, Chapter 6.
4. Walker, H., "Planning Tool Eases Shift To Precipitators During Scheduled Outage," *Electric World*, March 24, 1969.

EXERCISES

1. Develop an expression for the number of iterations required to process separately two networks that have mutual interfaces. The expression should be a function of the number of changes in direction among the interfaces.
2. Prepare a "cradle-to-grave" list of steps involving critical path methodology that would routinely be applied to potential new projects in an aerospace electronics firm or in a construction firm. Break up this list into stages such as bid-stage, initiation of work stage, work-in-process stage, and summarry performance on completed projects.

APPENDIX 12-1
DRAFT OF SPECIFICATIONS
FOR THE CRITICAL PATH METHOD
APPLIED TO CONSTRUCTION
PROJECTS

The following is a draft of specifications for the use of the Critical Path Method as a means of (1) insuring that a thorough and feasible plan for the project is developed, and (2) providing a form for detailed progress reports throughout the project. In the use of CPM as a contractual requirement, it is important that the Owner be represented by a person trained in the review of CPM networks and progress reports. The wording contained in this draft is intended only as a guide, to be modified or supplemented to achieve the reporting purposes desired for particular projects. This draft is concerned with schedule progress only, and does not include requirements for reporting the value of work in place.

PROJECT SCHEDULING AND PROGRESS REPORTING

(1) Within ___ calendar days of written notice of award of contract the Contractor shall submit to the (designated representative of the) Owner a network plan and a schedule of the job in the Critical Path Method (CPM) format. The basic diagramming method, computational procedures, and terms of CPM are contained in Chapters ___ through ___ of Reference 1.

(2) The CPM network may utilize either the arrow or node system. The network must contain detailed representation of all significant aspects of the construction plan, including but not restricted to site preparation, structural work, interior and exterior finishing, electrical and mechanical work, and acquisition and installation of special equipment and materials. For all equipment and materials fabricated or supplied especially for this project, the network shall show a sequence of activities, including preparation of shop drawings, approval of drawings, fabrication, delivery, and installation. The network shall contain between ___ and ___ activities, excluding necessary dummy activities. Each activity shall be identified on the network by a brief description and shall be assigned a single time estimate in working (or calendar) days.

(3) The initial and subsequent CPM schedules shall include the early and late start times, early and late finish times, and the total slack for each activity. The initial schedule shall indicate an early completion date for the project that is no later than the project's required completion date. For the schedule computations, either manual or computer methods may be used; in either case the notations and abbreviations used shall be fully explained. If the activity early and late times are not given in calendar date form, a conversion table shall be provided that will enable the Owner to convert the activity times into calendar dates. ___ complete copies of the initial network and schedule shall be submitted.

(4) While recognizing the Contractor's responsibility to determine the sequence and time estimates of the project activities, the Owner reserves the right to require the Contractor to modify any portion of the schedule judged impractical, infeasible, or unreasonable. Schedules returned to the Contractor for revision or correction shall be resubmitted to the Owner for approval within ___ calendar days.

(5) Within ___ calendar days after approval of the initial network and schedule, the Contractor shall submit an updated schedule,

which includes all the actual activity start and finish dates and is based on a complete CPM computation. Updated schedules shall then be submitted every other Friday (or by the 10th of each month) thereafter. — copies of updated schedules shall be submitted.

(6) In order to realistically represent the schedule of remaining work, in updating computations the effective date of the computation shall replace the earliest dates of the first uncompleted activity in each chain if (a) the early start date of an unstarted activity is earlier than the effective date, or (b) the early finish date of an unfinished activity is earlier than the effective date.

(7) The Contractor shall review and amend the network as necessary each reporting period in order to realistically represent the construction plan, and shall submit ___ copies of a list of all activities added, deleted, or changed since the previous period. Upon request, the Contractor shall also submit ___ copies of the network, including all revisions to date.

(8) If a schedule computation results in a negative slack (behind schedule) condition, a written statement must accompany the schedule, explaining the cause of the negative slack and the action planned by the Contractor to bring the project back on schedule.

(The following paragraph may be added for large projects.)

(3A) The CPM schedule shall consist of a list of the project activities sorted and tabulated in the following ways:

(a) by predecessor event number as a major sort and by successor event number as a minor sort.

(b) by total slack, from the least to the most, and

(c) by late start date, in chronological order.

Actual start and finish dates should be indicated for each activity that has started or finished. In sorts (b) and (c) dummies and finished activities may be omitted. (Other sorts may be desired, depending on the project and the network technique employed.)

SOLUTIONS TO EXERCISES

Chapter 2

1.

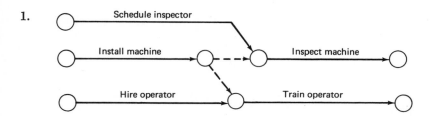

3.

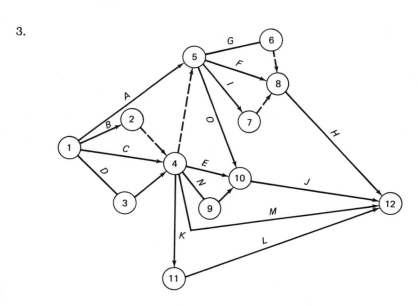

5. These are resource dependencies, indicating that there is only one crew of carpenters for the forming work. The dashed arrows from events 32, 37, and 42 also show that some subsequent activities are dependent on the use of forms that are removed in activities 31–32, 36–37, and 41–42. (The subsequent activities are not shown.) In this case the resource is the reusable forms.

7.

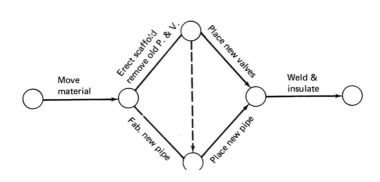

9.

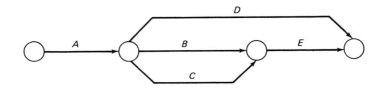

Chapter 3

1. Dig C does not depend on Pour A. The network has a false dependency.
3. This is a portion of a network for a structural concrete building. Shown is the second floor structural work, which is divided into three chains of activities corresponding to the three pours of concrete, labeled sections A-2, B-2, and C-2.
5.

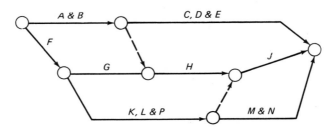

Chapter 4

1. a. No.
   b. 2.
   c. As late as possible, i.e., 6-7.
3. a. Critical Path 210-106-109   Slack = −2.
   b. No effect for a time of 16; however, a scheduled time of 12 for event

106 would make $L_{106} = 12$, and thus the path 210-106 would have a new slack value of -4.

5. c. Latest allowable start time for the deactivation of the lines is 205 hours.
    d. Activities on the critical path are A-B-F-G-M-D-O.
7. a. Project is one day behind schedule. Path 3-4-5-8 has a slack of -1.

Chapter 6

5.

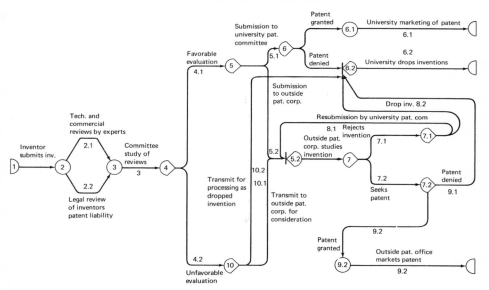

University patent committee policy procedure

Chapter 8

1. a. Yes

    Sequence of use:

Activity	Start	Finish
0–3	0	2
0–1	2	4
2–5	8	9
7–8	10	15

    b. 0-3, 4-5, 7-8, 16 Day Project Duration.

c.

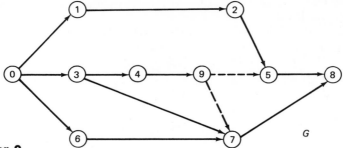

G

Chapter 9

1.

Project Duration	Direct Costs		Indirect Costs	Total Costs	Network Changes
12	610		900	1510	—
11	610+40	= 650	820	1470	D ↓ 4
10	650+40	= 690	740	1430	D ↓ 3
9	690+80	= 770	700	1470	A ↓ 2, D ↑ 4, G ↓ 3
8	770+100	= 870	660	1530	F ↓ 6, G ↓ 2
7	870+130	= 1000	620	1620	B ↓ 5, D ↓ 3, F ↓ 5

3. Parts a, b and d not given.

c.

Step No.	Project Duration	Direct Cost	Augmentation
"Normal"	382	1872	—
1	295	1888	105 ↓ 157
2	267	1898	105 ↓ 129; 110 ↓ 172
3	266	1899	105 ↓ 128; 110 ↓ 171
4	258	1903	105 ↓ 120 ; 110 ↓ 163; 109 ↓ 212
5	257	1905	110 ↓ 162; 109 ↓ 211; 131 ↓ 3
6	254	1913	110 ↓ 159; 109 ↓ 208; 102 ↓ 3
7	253	1916	133 ↓ 3
8	252	1921	110 ↓ 158; 109 ↓ 207; 112 ↓ 1
9	250	1934	110 ↓ 156; 109 ↓ 205; 120 ↓ 3
10	248	1949	110 ↓ 154; 109 ↓ 203; 124 ↓ 6

5. Activity i–j.

Minimize $(C_1 - C_2)\ \delta_1 - sy'_{ij}$
Subject to:

$$d_1\delta_1 + d_2\ (1 - \delta_1) + y'_{ij} + T_i - T_j \leqq 0$$
$$\delta_1 = 0 \text{ or } 1$$
$$0 \leqq y'_{ij} \leqq (d_3 - d_2)$$
$$(d_3 - d_2)\ (\delta_1 - 1) + y'_{ij} \leqq 0$$

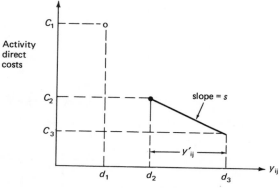

Minimize $(C_2 - C_3)\ \delta_2 - sy'_{ij}$
Subject to:

$$d_1 + y'_{ij} + (d_2 - d_3)\ (\delta_2 - 1) + T_i - T_j \leqq 0$$
$$\delta_2 = \text{non-negative integer}$$
$$-\delta_3 \leqq (\delta_2 - 1)$$
$$(d_2 - d_1)\ \delta_3 \leqq y'_{ij}$$
$$\delta_3 = 0 \text{ or } 1$$
$$0 \leqq y'_{ij} \leqq (d_2 - d_1)$$

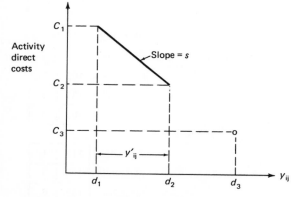

Minimize $(C_2 - C_3)$ $\quad \delta_2 - s_1 y'_{ij} - s_2 y''_{ij}$

Subject to:

$$d_1 + y'_{ij} + (d_2 - d_3)(\delta_2 - 1) + y''_{ij} + T_i - T_j \leq 0$$
$$(d_4 - d_3)(\delta_2 - 1) + y''_{ij} \leq 0$$
$$-\delta_1 \leq (\delta_2 - 1)$$
$$(d_2 - d_1)\delta_1 \leq y'_{ij}$$
$$\delta_2 = 0 \text{ or } 1$$
$$0 \leq y'_{ij} \leq (d_2 - d_1)$$
$$0 \leq y''_{ij} \leq (d_4 - d_3)$$
$$\delta_1 = \text{non-negative integer}$$

Activity duration

7. First construct pairs of pseudo activities for activity 1-2 and activity 1-3 as follows:

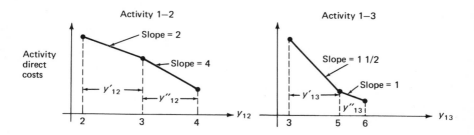

Maximize $\sum_i \sum_j c_{ij}y_{ij} = 2y'_{12} + 4y''_{12} + 1.5y'_{13} + y''_{13} + 2y_{34}$

Subject to
$$T_1 + y'_{12} + y''_{12} + 2 - T_2 \leqq 0$$
$$T_1 + y'_{13} + y''_{13} + 3 - T_3 \leqq 0$$
$$T_2 + 3 - T_3 \leqq 0$$
$$T_2 + 2 - T_4 \leqq 0$$
$$T_3 + y_{34} - T_4 \leqq 0$$
$$T_4 - T_1 \leqq 10$$
$$0 \leqq y''_{12} \leqq 1$$
$$0 \leqq y'_{12} \leqq 1$$
$$y''_{12} \leqq \delta_{12} \leqq y'_{12}$$
$$\delta_{12} = \text{non-negative integer}$$
$$0 \leqq y''_{13} \leqq 1$$
$$0 \leqq y'_{13} \leqq 2$$
$$y_{34} = 5\delta'_{34} + 2\delta_{34}$$
$$\delta'_{34} + \delta_{34} = 1$$
$$\delta_{34} = \text{non-negative integer}$$
$$\delta'_{34} = \text{non-negative integer}$$

Chapter 11

3. a.

Activity	a	m	b	t_e	V_t
1–2	1	2	4	2-1/6	0.8789
2–3	3	5	7	5	1.5625
2–4	1	2	3	2	0.3906
4–3	1	3	5	3	1.5625

  b. $7\frac{1}{6}$

  c. 2.8320

  d. $P\{T \leq 8\} = 0.69$; $P\{T \leq 9\} = 0.86$

  e. $9.946 \cong 10$

5. $P\{T \leq 16 \mid T_7 = 10\} = P\{Z \leq (16\text{-}15)/\sqrt{0.88}\} = 0.86$

9. a. The activities that have two or more predecessor activities are 108, 113, 117, 119, 121, 125, 126, 127, 128, 129, 132, and 133. The predecessor events of these activities are the ones where a merge event bias may occur. Of these 12 cases, only activities 121, 125 and 126 will present merge event bias problems. For the others, the expected means of the merging paths are quite far apart, or else there is a great deal of correlation in the two paths, so that the bias will again be negligible, e.g., activity 132 and 129.

     It should also be noted that the merge event biases will occur on slack paths, and hence will not affect the critical path, and hence the conventional PERT analysis should be a good approximation in this example.

  b. The critical path is made up of activities 102, 105, 108, 112, 115, 118, 120, 123, 124, 128, 131, 132 and 133. The sum of the estimated mean times for this path is 382. The variance of this path can be approximated by considering only activities 105 and 118. The sum of their variances is 1221.3 and the approximate standard deviation is 35.

  c. $P\{T \leq 400\} = P\{Z \leq (400 - 382)/35\} = 0.70$.

INDEX

INDEX